EINSTEIN & THE EVOLUTION OF QUANTUM PHYSICS

FROM A LIGHT BEAM TO DEEP COSMOS

DEV ARASTU PANCHARIYA

HESTEN
an imprint of BlueRose Publishers

BLUEROSE PUBLISHERS
India | U.K.

Published under Hesten, a traditional imprint of BlueRose

For permissions requests or inquiries regarding this publication, please contact:

BLUEROSE PUBLISHERS
www.BlueRoseONE.com
info@bluerosepublishers.com
+91 8882 898 898
+4407342408967

ISBN: 978-93-5668-246-7

Cover design: Shivam
Typesetting: Namrata Saini

A Tribute To...

Foreword

Indian Institute of Technology Jodhpur
NH-65, Nagaur Road, Karwar, Jodhpur 342037, INDIA
Phone: (0291) 280 1006; eMail: srv@iitj.ac.in, svadera59@gmail.com

Dr. Sampat Raj Vadera
Deputy Director &
Professor, Department of Physics
Formerly, Director, DRDO, Jodhpur

On October 09, 2023, I was to deliver a Keynote Lecture in an International Conference on Condensed Matter and Applied Physics. During the lunch break, I was introduced to a young gentleman, named Dev Arastu Panchariya, as a successor to Albert Einstein. I got engaged into deep conversation with him it continued for nearly about 2 hours. During the conversation, he could generate a feeling as if we had known each other for years.

During our interactions, I could see him venturing into a wide range of topics with great clarity and originality of thoughts. I was deeply impressed by his intellect, knowledge, attitude and particularly would like to appreciate the vision he has at such a young age.

I am particularly happy to know that he has written an excellent book entitled "Einstein & the Evolution of Quantum Physics: from a light beam to deep cosmos" describing evolution of a highly important area which has greatly influenced the developments in science and technology for more than a century and is also currently shaping as one of the most emerging areas which is expected to impact human life very significantly in times to come. I am sure his book will hugely benefit not only the younger generation but will also greatly enrich the knowledge of the experienced academicians and scientists across the world.

I take this opportunity to compliment Mr. Dev Arastu Panchariya for putting his thoughts together in a wonderful creation namely "**Einstein & the Evolution of Quantum Physics: from a light beam to deep cosmos**".

I wish him great success in all his future endeavours.

(Sampat Raj Vadera)

Special Prologue

True power is not just attainment of knowledge but the ability to apply and share it in a way that humanity can benefit from it. The Book "Einstein & The Evolution of Quantum Physics : From a light beam to Deep Cosmos" is a testament that Dev Arastu Panchariya possesses this extraordinary power. With my first meeting with Dev Arastu, I could sense a rare and exclusive variety of energy, which was never experienced before and I'm deeply grateful to the destiny and key Ministerial and Honorable Parliamentarian figures as the way for my first interaction with a great mind like him.

I strongly believe that with the times to come, Dev Arastu's intellectual power, perspective, and contribution to the world of wisdom at large will keep on integrating with greater dimensions.

I am sure that this book will create bold enthusiasm among the youth and across generations about the deep cosmic phenomena and the discovered and resilient possibilities of science and philosophy.

Deepak Ojha

Managing Director,

Waldies Compound Limited

Table of Contents

Introduction

No observer, no universe. It is natural for one to ask how that is even possible. How can the sun, the moon, stars and galaxies, etc. suddenly disappear into nothing if there is no observer looking at them? It doesn't seem to make any sense. Welcome to the universe of quantum physics. It doesn't have to make perfect sense for it to be true.

Albert Einstein, one of the pioneers and founding fathers of quantum physics, agreed that it worked perfectly but he still considered it as spooky action from a distance and maintained that it was an incomplete explanation of reality. He once famously asked his friend and physicist Abraham Pais, if he really believed that the moon exists only when he looks at it.

Einstein's view regarding the universe was deterministic where God does not play dice with the universe. However, in the quantum realm it seems like God does indeed play dice and the behaviour of subatomic particles is nothing short of miraculous, where the real and virtual mix to create matter itself. The most fundamental properties of particles like location, momentum and energy are all probabilistic and cannot be determined precisely simultaneously. In other words, quantum physics is all about probabilities and the quantum world view is probabilistic.

The earliest record of subatomic particles predates 6th century BCE. But it wasn't until late 19th and early 20th century that subatomic particles were discovered and a mathematical framework was laid for quantum mechanics by Scientists like Einstein, Max Planck, N

Bohr, Schrödinger, P A M Dirac, Louis de Broglie, W Pauli, Max Born, W Heisenberg, Fermi, Bohm, etc. Their contributions were taken further by some of the most brilliant modern scientists including R P Feynman, J S Bell, J Schwinger, etc.

Unlike the account of classical physics, there is no unanimous single interpretation of quantum mechanics. Even after almost a century, the Copenhagen Interpretation by Niels Bohr and Werner Heisenberg is still considered the gold standard. Due to wave function collapse, the universe is at its most probabilistic and not deterministic.

There are mathematical possibilities of several other interpretations while the true interpretation of quantum mechanics is still up for discussion. Its practical applications are very precise and have revolutionized almost every scientific field, like smartphones, computers, internet, telecommunications, GPS, quantum computing, lasers, atomic-clocks, quantum cryptography, MRI, LED, quantum biology, quantum chemistry, etc.

Some argued that in modern times, quantum physics seems to have some deeply rooted connection with metaphysics too, i.e. consciousness in particular or other related abstracts.

"Light" or "Quanta" was somehow a starting point of this quantum world, but 'how story came up to Quanta', 'what about the initial point of the conceptualization of light' and so on, are some of the radical questions people usually don't even ask or think of. Apart from the subsets of the quantum world, like statistical mechanics and so on, there is this intersection of quantum and particle physics i.e. Quantum Field

Theories, consisting of very fascinating aspects like the famous example of Higgs Boson. This intersection has put forth many great forms of sciences that have something to do with famous notions including the Big Bang, quantum fluctuations, string theory in terms of going beyond elementary particles and compositions like quarks & hadrons respectively.

The rise of quantum field theories, especially quantum electrodynamics even provided the compatibility of quantum world with the Special Theory of Relativity. Quantum physics as one of the cores of theoretical or mathematical Physics has a great hand in exploring the idea of Theory of Everything as the unified field theory, the dream of Einstein that is basically about the unification of all fundamental forces, although now only Einstein's General Relativity is yet to be unified as the most precise explanation of Gravity for which we have a hypothetical elementary particle i.e. Graviton regarding the phenomena of Quantum Gravity to be precise, however Loop Quantum Gravity and String Theory also stands as a probable candidates for the same, however with different approaches. Reading anything great about quantum physics generates its own excitement about many things described previously, including the EPR Paradox or quantum entanglement (which can be related with simultaneous action in simplified and general terms), hidden variables, mechanisms by S N Bose and Albert Einstein, or by Bohr and Einstein, etc. All the mentioned events are just briefs, but the full story has its own systematic unfolding, which is elaborated in this outstanding book that is right from the pre-Socratic times up to now and even to the future, this is an entirely new way to put forward the unfolding of quantum physics. The Genius of Dev Arastu has beautifully expressed and explored the whole account of quantum world which makes this book a unique entity and a lifetime experience. The book, I feel, fulfils its concern for both technical and non-technical audience because it contains some great advice and insights with a huge amount of content that can lead technical people to some revolutionary new paths and ignite a passion in the general audience with a great intuitive collection of quantum realm. This entire story of the quantum world has been expressed with such perfection and simplicity that it is also a feat in itself, and hence he is a noteworthy inspirational figure for everyone. I wish him more success in his future endeavours to take this quantum leap to a Unified Theory of Everything.

Dr. Ravindra Mangal

Professor Emeritus and Former Head, Dept. of Physics

Former Director of Research

M.G.S. University

President – Condensed Matter Research Society

India

Part 1:
A Brief Commencement of Quantum Physics

1
A Brief Overview of Quantum Physics

God does not play dice with the Universe.
- Albert Einstein

Universe is kind of a set of mysteries and certain extraordinary personalities in the history of time who have decoded multiple aspects in quite elegant ways and one of those is - "Quantum Physics", which is determined to decode the subtle nature of reality and thus quantum physics is fundamentally a kind of probabilistic way to unfold nature or to determine the abstracts at a very microscopic level. Basically, the birth of quantum mechanics that took place in the contemporary times is attributed to pioneers in the history of modern physics like Albert Einstein, Max Planck, Max Born, Wolfgang Pauli,

Werner Heisenberg, Erwin Schrödinger, etc. The term 'Quantum Mechanics' was introduced in a 1924 paper by Max Born, titled "Zur Quantenmechanik", which means "On Quantum Mechanics".

Quantum mechanics also predicts certain laws and aftermaths of classical mechanics but sometimes in a limited way, although the insights seem to be quite satisfactory and elegant in both theoretical & mathematical perspectives as it describes things in a very fundamental way. The relationship of quantum physics with the classical one is a way to figure out the corresponding solution in a generalized way. For example, this comprehension in a peculiar course of action gives rise to the concept of "Quantum Chaos", which in some sense relates or deals with the description of classical dynamical system as quantum theory, but this also develops the relative system in a chaotic way, which is of course some sort of constitutional description of quantum mechanics - which is precise too.

The intrinsic property of quantum physics is that it gives rise to the possibilities of certain systematic superposition. Or, more foundationally, it is a kind of systematic study of minuscule notions or particles, sub-atomic particles, matters, molecules, etc. and at more microscopic levels like atomic, nuclear, etc. and corresponding interactions.

It also deals with very fundamental energy levels of those different particles or other such forms. The laws & respective notions regarding quantum physics are considered as one of the most foundational explanations of natural abbreviations, which is also a very obvious aspect to conceive due to its mathematical elegance and analogous theories or concepts.

The dealing of this branch keeps having new applications and discernments in both the real world and theoretical as well with the passage of Time, and if it goes in the same or more effective modus operandi then this will develop several new and revolutionary insights in future with more precision and accuracy. For example, it has applications in following branches other than core Theoretical Physics:

- Computer Science

  Quantum computing is a step ahead in terms of technological aspects to perform discrete computations. It fundamentally applies the quantum physics' favoured phenomenon such as superposition, quantum amplitude and beyond, to determine or to precede the relative computations. The concept of such an abstract arose in early 1980s with a quantum mechanical-based model or solution to the legend, Alan Turing's "Turing Machine" led by an American Physicist Paul A. Benioff.

  Briefly, the Turing Machine is fundamentally a mathematical model to a kind of abstract machine, which performs the logical manipulation of data in a discrete level of artificial memory cell like a Central Processing Unit (CPU).

  This is one of the pioneering works of the mathematician, computer scientist, cryptanalyst and philosopher Alan Turing. Later on, another two prominent physicists, Richard Phillips Feynman and Yuri Manin proposed the concept of "Quantum Simulator" which suggests that a

Quantum Computer can simulate data or things which are not possible by the means of a classical computer and this concept has also raised another fundamental field began with a primordial problem, abbreviated as the "Hamiltonian Simulation", which refers to the findings of Quantum Algorithms required to simulate the things at certain discrete level.

- Cryptography

Cryptography or cryptology was first determined as cryptogram. But this term was first taken into account by an American writer, poet and editor named Edgar Allan Poe. This branch or field of study is considered to determine the secret codes of any system in a very logical and analytical way.

This is developed or applied with an intersection in betwixt various other fields like mathematics, physics, computer science, etc. In modern times, a concept called "Quantum Cryptography" is on the way to intensive advancement of sciences in a versatile form.

This field is basically a sort of implementation from quantum mechanical properties to scrutinize the cryptographic engagements, which is strictly attributed to the "Quantum Key Distribution".

Also, it uses certain techniques from quantum computing, referred to another subset of "Quantum Cryptography" that is "Post-Quantum Cryptography", attributing the quantum stated or levelled mathematical modelling or algorithmic inspects with respect to the quantum computing.

- Biology

  In the field of biology, various systems or system of interactions like formation of excited state of electron or shifting of higher energy levels, chemical reactions, transfer of required ions in chemical processes, light absorption, etc. are considered under the realm of "Quantum Biology". This analyses or determines the relative physical and chemical processes in terms of the laws of Quantum Mechanics with both theoretical and mathematical frameworks in order to explore the biological systems in a precise methodology.

  The trailblazers of quantum mechanics realized the correlation in betwixt these two systematic explanations of different processes. For example Niels Bohr, with his few eminent colleagues, described his "Principle of Complementarity" as a foundational aspect to the life sciences. His principle briefly says that all entities have ineluctable set of distinctions, which cannot be observed or measured concomitantly, for example wave-particle duality, position and momentum (as per Heisenberg's Uncertainty Principle that simply says the position and the velocity of an object cannot both be measured exactly, at the same time, even in theory) and Erwin Schrödinger, in his 1944 book "What is

Life?", explored the applications of quantum theory in biology.

- Some other such examples are atomic clocks, MRI (Magnetic Resonance Imaging), lasers, etc.

Quantum mechanics, instead of giving direct predictions of anything, gives possibilities with respect to the data or just determines the results in a definite likelihood prospect as per its fundamental explanation in both theoretical & mathematical realms; in the world of quantum physics, the entire universe is, for all intents & purposes, like a concatenation of possibilities or probabilities, which is a quite serious issue of the identification of the universal truth or unifying it with the ideas of Albert Einstein, as once he also said, "God never plays dice with the Universe".

But apart from this, quantum physics has its own beauty of defining relative things.

As time passed on, a few very eminent scientific people brought the concept of Consciousness into the account of quantum physics; for example, Eugene Wigner had an idea that quantum physics has something to do with the mechanism of mind and thus he, with another eminent Scientist John Von Neumann, gave a kind of modern interpretation in favour of this, which was: "Wave function collapses due to its interaction with Consciousness" or more precisely, the "Von Neumann-Wigner Interpretation". A lot of examples like this are available out there which suggest a kind of abstract touch of metaphysics in the account of quantum physics for instance, a distinguished Physicist Freeman Dyson stated that, "Mind, as manifested by the capacity to make choices, is to some extent inherent in every electron", and this also indicates in the same direction towards the unification of the quantum physics & Consciousness.

But here's a catch and my opinion on it, which is of course put forward with all due respect as I'm not opposing anyone's perception or rebelling against any interpretation that is currently in the account of theoretical physics or quantum physics.

There is a question corresponding to the stabilization of quantum mechanics and Consciousness: "Is Consciousness' course of action an emplacement of probabilistic chances or respective outcomes?" If the answer is no, then we should keep both separate and proceed with the study accordingly; but if the answer is yes, then if distance between two different places are similar to two other different places, it doesn't mean that both the places are identical. Hence we should still take them as separate cases because mixing up two different aspects together never makes great contributions to humanity. But yes, they can be good together if we are intent on having half or a single facet of truth.

2

Origin of Quantum Physics

Music is a moral law. It gives soul to the universe, wings to the mind, flight to the imagination, and charm and gaiety to life and to everything. – Plato

The origin of quantum physics is certainly quite fascinating and a different kind of phenomena, like storytelling with ups and downs. Though Ludwig Boltzmann was the scientist who postulated or propounded the discrete behaviour of the energy levels of physical system like molecules in 1877, and his profound work and investigation became the greatest reason behind the establishment of statistical thermodynamics and statistical mechanics. But, apart from this, a lot of other aspects are there which are further discussed in the upcoming parts.

But the account of this current "Quantum Physics" that we're dealing with since several decades starts with the black-body paper determined in year 1900 from the attribution of one of the greatest scientists of all time, Max Planck. But the black-body paper itself has a background story with tied intricacies in history of philosophy that has been a true quest into the chapters of Quantum story and is now contributing to the large-scaled Universe too.

After that, around 1905, an incredibly gifted legend with true genius had an entirely original idea, taking Planck's idea further with a fascinating insight which used Planck's black-body radiation to theorize a revolutionary concept, which later earned him a Nobel Prize for "his services to Theoretical Physics, and especially for his discovery of the law of the photoelectric effect" in 1921. But by then he already turned over the historic spectrum. And the name of that great persona is Albert Einstein. Although later on, when other scientists came up with their ideas to build this new branch of Physics onto a next level, Einstein wasn't fully satisfied with what was happening in the realm of quantum physics because it had started to contradict the deterministic nature of the universe and so, according to Einstein, it required certain fundamental modifications.

Then, after this comes another legacy of Einstein - his theoretical and mathematical contributions over the "Brownian Motion Theory" described by a botanist Robert Brown in 1827, which took place at the time when he was looking through a microscope at pollens of the plant *Clarkia Pulchella* immersed in water. Einstein's development of this paper not only contributed to an entirely revolutionary aspect of quantum physics but was also responsible for a major and modern headway to statistical mechanics to analyse various quantum methodologies with more precision. That was the time when something known as "Quantum Physics" came into existence. While the Theory of Relativity was developed by a single person, this systematic mechanism for the study of microscopic level of abstracts took several great minds to evolve as a thoroughly new branch and one of the most significant pillars of modern physics.

Einstein certainly made a lot of major scientific contributions in the evolution of quantum physics into the account of theoretical physics which are discussed in greater detail in one of the later chapters that focus on Einstein's key contributions to the minuscule world of subtle mysteries.

Although Einstein was always very clear about the deterministic nature of the Universe and always believed it too, he wasn't entirely satisfied with the explanations of quantum physics, so much so that Einstein always imagined this new branch to theoretical physics in his own regime of determinism, and of course his way of rational and intuitive thinking, which lead to his contributions to the field and absolutely made him one of the founding fathers of quantum physics.

Due to the distinctive approach towards the universe, Einstein's letter in 1926 expressed that "Quantum mechanics is certainly imposing. But an 'inner voice' tells me that it is not yet the real thing. The theory says a lot, but does not really bring us any closer to the secret of the 'old one'. I, at any rate, am convinced that He does not throw dice".

This letter was written to another brilliant scientist, Max Born who in some sense believed that "God plays dice" or vice-versa from Einstein. He also made several valuable contributions in the corresponding fields like Born Rigidity, Matrix Algebra, Born—Haber Cycle, Born Series, Born–von Karman boundary conditions, Born—Landé equation, etc., and he was also the man who explicated the Schrödinger's wave interpretation as a probabilistic amplitude, which indicates that the wave must be squared to obtain the indispensable probability or, in other words, he postulated that squaring of the wave function doesn't give rise to the physical density of electron waves but just their probability density and this postulation is also widely known as Born Rule or Born Law; it was theorized in the same year as the letter correspondence with Albert Einstein, around 1926.

This interpretation of Max Born clearly shows that his interpretation of the reality was full of 'possibilities' in nature, which Einstein obviously never accepted in relation to the account of quantum physics.

3
Underlying And Initial Trials With Reality

The nature of light is a subject of no material importance to the concerns of life or to the practice of the arts, but it is in many other respects extremely interesting.
– Thomas Young

In this new field of study, due to the greatest scientific pioneers of all times, it also had certain peculiar experimental aspects to deal with in order to relate the Reality with its theoretical and mathematical explanations. Most of the things were under the attribution of "Modern Quantum Mechanics" or the quantum theory after or in 1925 when other eminent scientists like Niels Bohr, Max Born, Erwin Schrödinger, Louis de Broglie, John von

Neumann, Werner Heisenberg, Wolfgang Ernst Pauli, Arthur Holly Compton, etc.

A lot of relative things are discussed in the further portion though, carrying all the modern interpretations simultaneously. However, for now, let's understand in brief how the scenario of quantum mechanics deals with the reality of Nature and for the sake of this we have to take the famous Double-slit Experiment into consideration, which was first performed by Thomas Young in 1801. When scientists came as the foundational pillars of modern physics, they took the double-slit experiment or Young's interference experiment into consideration in order to search for truth or to seek reality with respect to nature, and hence the experiment was on the rise for evidence related aspects of different methodologies or theoretical and mathematical abstracts by different scientific people, and the observations from that entirely favoured the concept of superposition and indicated that the scientific reality, which legends like Einstein and Newton looked for, was still certain steps ahead in those days.

Apart from that, one more attestation that favoured the double-slit experiment emerged with the wave theory of light. Firstly, Thomas Young, who is one of the pioneers of the wave theory of light, says that light is also somewhat of a wave motion. He considered his interpretation for light to be more profound as compared to René Descartes' concept of corpuscular theory of light, which suggested Light as a formation of discrete particles with a certain velocity subject to limitation and travelling in a straight line. This concept was also taken further by the pioneer of several physical and mathematical theories, Isaac Newton. Although the classical interpretation for the behaviour of light had peculiar limitations, which were considered with a more generalized view over the corpuscular theory of light, proposed in 1678 by Christiaan Huygens, almost 6 years after Newton's adoption of the theory. Huygens first described this generalized perception on the theory nearly after 12 years, in1690, in his "Treatise on Light", which is considered as the first ever mathematical interpretation over the theory of light. But we were at the double-slit experiment.

So, in short, Thomas Young demonstrated it in order to favour the wave behaviour of light. This experiment was a part of classical physics until the modern exegesis came into the existence i.e. quantum mechanics and specifically the wave - particle duality, which advocates all quantum entities, may be illustrated as either a wave or a particle (or both wave-like and particle-like characteristics). This developed another derivative as the ineptitude of classical expounding up to the microscopic level or quantum scale, and almost simultaneously Louis de Broglie proposed that any kind of discrete configurations of matter have both wave and particle-like behaviour, which was an answer to Einstein's interpretation on photons because after this idea of Einstein, the question was about the wave-like behaviour of light – in other words, is this only true for light or for the material objects too? The de Broglie Hypothesis also shows that wave-particle duality is not only an abbreviation to light but a fundamentally manifested property of both radiation and matter, and hence this hypothesis also worked as a great hand in the development of the theory of atomic structure and particle physics in unison with the quantum mechanics.

Now, in 1927, American physicists Clinton Davisson and Lester Germer verified the de Broglie Hypothesis with their "Davisson-Germer experiment" at the Bell Labs where they fired electrons at a crystalline nickel target and the resulting diffraction patterns with interference matched with the predictions of de Broglie Hypothesis of wavelength; in the year 1929, de Broglie, notably became the first scientist ever to win a Nobel Prize for his doctoral research. Also, an English physicist named George Paget Thomson also demonstrated a similar kind of experiment by firing electrons through metal films, which produced a similar kind of diffraction pattern. For their respective contributions, both Clinton Davisson and George Paget Thomson were jointly awarded the Nobel Prize in 1937 "for their experimental discovery of the diffraction of electrons by crystals". But the main catch of the story is the partial completeness of the quantum theory towards the mysterious reality. The respective experiments in quantum physics corresponding to the theories, hypotheses or mathematical formulations have a certain level of approach to the natural reality, but it has always been mysterious or half complete according to a lot of great scientists, especially Albert Einstein. But yes, it's quite elegant and there is no question about that.

4
The Ontological Epistemology

When the province of physical theory was extended to encompass microscopic phenomena through the creation of quantum mechanics, the concept of consciousness came to the fore again. It was not possible to formulate the laws of quantum mechanics in a fully consistent way without reference to the consciousness. – Eugene Wigner

The area of quantum physics is becoming exclusive and the metaphysical perceptions seems to have a special place in its consideration for some accounts. For example, the unification of quantum mechanics with Consciousness as previously defined, a few intellectuals regard this unification of two different regimes as quite a significant enosis of Science & Metaphysics. However, the rationalism for this regime is definitely compromised from a scientific viewpoint. There is obviously no question on the precision and accuracy of quantum mechanics to relate the theoretical, mathematical and experimental frameworks, but the truth remains as it is, which is it's intrinsic characteristic in order to observe Nature with peculiar uncertainties and probabilities. We know the internal nature of quantum physics very well, but we don't know about the current classifications over metaphysical reference in its corresponding field of study in Philosophy.

So, how do we relate both these realms in the same manifestation? I'm definitely not being critical of quantum mechanics and consciousness or other such unifications; I'm only just offering my opinion with respect to this concept which I suppose are rational too as described previously in one of the parts of metaphysical interpretations pursuing themselves in to the universal notions.

Also, there is a tendency favouring the integration of Science and Consciousness, called the "Quantum Mind" or "Quantum Consciousness", which says that classical mechanics can't explain Consciousness and have no clarifications regarding the functioning of the human mind but quantum mechanics on the other hand has that capability. This new interpretation originated and was favoured by the philosophy of a few pioneers of quantum mechanics, but a few scientists also never agreed with it. For example, once the American particle physicist and philosopher Victor John Stenger classified "Quantum Mind" as a myth which was being considered into quantum world "without any scientific basis" at all.

Well, according to my perception, 'myth' is a very strong word but we should certainly focus on the answer of previously asked questions and one more – is

Consciousness or Nature or Universe a probabilistic Game? So it can be run with the probabilistic predictions. Although its previously mentioned that Quantum Physics works in its own way with respect to its theoretical, mathematical and experimental frameworks to define certain aspects, and quantum mechanics has also been a very successful tool to explore the abstracts, which were undiscovered since they came into existence, hence made a revolutionary contribution to sciences that promoted humanity in a beautiful and significant perspective. But still, defining the entire Universe as a whole is a big deal. Perhaps it will be possible someday, albeit with certain modifications in the current quantum physics as described by Einstein.

Apart from these facts regarding consciousness, there are certain other aspects which are interpreted by many others that quantum mechanics has something minutely to do with the philosophy of metaphysics, however, analyzing a few of them deeply might encounter as rationally equipped. A few of them are mentioned below:

- Quantum Mysticism

Quantum mysticism is a kind of metaphysical involvement (also referred to as Consciousness) into the quantum considerations that give multiple insights in form of quantum mechanics' interpretations by various eminent scientists of both early and modern times. But on the other hand, Albert Einstein always objected to such kind of interpretation. Most effectively, Erwin Schrödinger and Werner Heisenberg were the two physicists who strictly believed in that kind of explanation or correlation between quantum mechanics and metaphysical aspects and thus, also justified a few of their works on that basis. But there were a few scientists who called this mixture of the two domains a 'pseudoscience' (beliefs to be claimed as both scientific and factual but have nothing to do with the core sciences).

- Quantum Mysticism with Neuroscience

At the same time, there are a few principles or interpretations of quantum mechanics that indicate the essence of the metaphysical aspects in the core theories. For example, "Holonomic Brain Theory" developed by American neuroscientist Karl H Pribram with a significant attribution to another pioneer of quantum theory, David Bohm, who is best known for his many contributions to the field such as the de Broglie—Bohm interpretation, Bohm diffusion, etc. Briefly, this "Holonomic Brain Theory" says that the Consciousness as a subset of human mind is formed by quantum effects. But, if we go through some other works of David Bohm in line with this perception, he is favouring determinism in form of the "Hidden Variable Theory", which is fundamentally regarded as the allocation of the deterministic explanations to the phenomena of quantum mechanics. So here also it can be observed that the work by a single person is creating two different pictures of reality.

Also, there is another term in Modern Neuroscience with respect to quantum physics or more precisely the Quantum Field Theory (this theory is basically concerned with the particle interactions or such systems on the basis of physical models specifically focused to deal three pillars of theoretical physics simultaneously, i.e. Classical Field Theory, Quantum Mechanics and Special Theory of Relativity up to a certain level; they don't allow Einstein's General Theory of Relativity to be included in the regime).

This term is known as "Quantum Brain Dynamics", which is theorized to determine the functioning of brain in terms of (or with the help of) quantum field theory. But some modern scientists also raise the question whether the theory can be used to analyse the system of Consciousness through this idea, because quantum physics is the only current theory that relates superposition and consciousness.

- Von Neumann—Wigner Interpretation

The Von Neumann—Wigner Interpretation was one of the origins of this metaphysical account in respect of quantum mechanics. The story of this starts with another interpretation known as "Copenhagen Interpretation" which has two great hands of quantum mechanics, Werner Heisenberg and Niels Bohr. In between 1925-1927, as is described in some of the upcoming parts, the quantum story systematically evolves into chaos. This interpretation is simply suggesting the truth of quantum theory, that it only gives a probabilistic view of a given measurement's possible results or experimental one. Although this interpretation has a lot of different causalities and consequences which are quite fascinating, but a quite significant aftermath of Copenhagen interpretation linked to our current point of discussion is called 'Wave Function Collapse' as it gave birth to another very important interpretation - "Von Neumann—Wigner Interpretation", which is attributed to two more leading scientists - John Von Neumann and Eugene Wigner. In this interpretation, sometimes, its interpreted that consciousness (with reference to a particular observer) is theorized as a fundamental prerequisite in the procedure of quantum measurements and also indicates that consciousness causes the collapse of wave function (wave function is one of the most noteworthy part of Wave Mechanics - a mathematical explanation to any isolated quantum system with certain properties like its responsibility for the establishment of probability distribution in three dimensions etc.). Now, there is another story regarding the legacy of another pioneer of quantum theory, Erwin Schrödinger i.e. the "Schrödinger's cat Experiment". This 'Cat Experiment' is a paradox or thought experiment contrived in 1935. The idea had the great hand of Albert Einstein in it as well. This indicates a blank in the Copenhagen Interpretation if applied in the everyday life and finally theorize a hypothetical cat to be simultaneously dead or alive. But in 1961, Eugene Wigner redefined the Schrödinger's Cat thought-experiment as "Wigner's Friend", which ultimately postulated that the presence of a conscious observer is must for anything to be detected as in the case of quantum measurements and this abstract also helped the 'Von Neumann—Wigner Interpretation' to evolve in a more profound way.

This interpretation is also favoured by certain mathematical aspects, which are again based on the foundation of superposition in the quantum state (this term or 'Quantum State' refers to an isolated quantum system that provides a probability distribution for each and every outcome in quantum measurements. This is also categorized in two different forms i.e. 'pure quantum state' and 'mixed quantum state'). Hence, the 'Von Neumann—Wigner Interpretation' also relates or provides a strong linkage between Consciousness and Superposition. The interpretation favoured both Werner Heisenberg and Niels Bohr as Bohr always thought that quantum physics can explain everything and that it provides a suitable description for nature and took active interest in the corresponding philosophy regarding quantum physics.

If we observe deeply, perceptions or formulations of a lot of great scientists and polymaths don't intersect anywhere close with the entirety of concept of Quantum Mysticism and its respective subsets, this possibly caused due to irrationality involved in metaphysical identifications. These minds included René Descartes, Isaac Newton, Albert Einstein, Christiaan Huygens, etc. But those modern scientists who were close to the era of the development of quantum physics (like Albert Einstein) never entirely denied its significance, but just appealed for certain modifications and for excluding much metaphysical aspects from its core, which I support too. For example, when we have some postulation like "Electromagnetic Field Theories are also responsible for the study of Consciousness", then the question that comes to the mind is why don't we take both concepts further on individual basis; perhaps up to some extent, Einstein also had such collective facts and questions on certain aspects regarding the explanations in quantum physics and beyond.

Part 2 :
Foundational Insights of Scientists Regarding the Ephemeral Antiquity of Light

5
The Ephemeral Antiquity of Light

Heat can never pass from a colder to a warmer body without some other change, connected therewith, occurring at the same time. - Rudolf Clausis

Who knew that a single "Thought ingredient" would become one of the most explored and curious branches or fields of study in the physical consideration. Up to the time when quantum physics wasn't in the account of theoretical and experimental physics, there was only one aspect which made scientists think over centuries and the result of that intense thinking was the foundation of this new branch of physics - Quantum Mechanics; that one aspect or entity that made many eminent and brilliant scientists and polymaths contemplate over was Light. This is the point where going through different eras or decades, scientists felt some profound explanation in its favour and that vision towards its explanation was somehow the greatest reason towards the emergence of another significant branch of physics in form of the revolutionary insights which came to be called Quantum Revolution. This sole segment known as Light has a lot of different interpretations in form of several philosophies, theories, concepts, hypotheses, experiments, thought experiments, mathematical expressions, etc., lead by some of the greatest legends of the universe of Science and Philosophy such as Empedocles, Titus Lucretius Carus, Aristotle, Socrates, Democritus, Epicurus, Euclid, Claudius Ptolemy, Galileo Galilei, Johannes Kepler, Pierre Gassendi, René Descartes, Isaac Newton, Robert Hooke, Christiaan Huygens, Leonard Euler, Thomas Young, Albert Einstein, Max Planck, etc. The respective legacies of a few historical minds that played a vital role in the evolution of the theory of Light enfolds in the following Chapters.

6
Empedocles

***None* of the *gods has formed* the *world, nor has any man,* it *has* always been. - Empedocles**

Empedocles was a pre-Socratic Greek Philosopher and being one of the greatest ever in the history of time who intensively contributed in various ways to the study, including advanced concepts which are currently quite active in sciences like Cosmogony, Ontology, etc. And of course, he was one of the fathers of these terms and respective methodologies. He was such a versatile person that it is even said that Aristotle considered him the inventor of Rhetoric (rhetoric is the theory that refers to magniloquence as the art of using language practised to coax others). The life of Empedocles was just as fascinating as his work and insights. The most fundamental facts about his life are debatable, as there is a huge uncertainty about his birth and death dates due to the lack of any attested source of information about it. Although, it's most commonly and historically believed that according to Aristotle, who was also regarded as one of the greatest Philosophers and Polymaths of all time, the life span of Empedocles was around sixty years. Apart from that, the questions over the respective dates lead a great French poet and Philosopher, Johann Christian Friedrich Hölderlin (1770-1843) to write a drama on the last days of Empedocles' life named as "The Death of Empedocles". This drama also remained unfinished. His deep imagination was also highly influenced by some of the legends of the field including the mathematician and philosopher Pythagoras and his corresponding philosophy abbreviated as Pythagoreanism, which originated in the 6th century BC.

Empedocles had many revolutionary insights and philosophies to contribute among which, a few were perhaps not as true as compared to the current explanations or frameworks of modern sciences, but they became the foundation to various perceptions of several scientific personalities which has existed after him including Aristotle, Socrates, etc.; this indicates the sagacity and beauty of his ideas. Although a few of his ideas became highly influential for the advanced times too, like his concept of four fundamental elements as "roots" i.e. earth, air, fire and water. By the time, this ideology had already been adopted by both Plato and Aristotle - among whom Plato is said to be the first person who used the word 'element' of the theory of four fundamental entities. Empedocles is also well-known for his idea of two cosmic forces that played a vital role in order to develop some elegant legacies with respect to some of great philosophers of the age after him. These forces are attributed as Love (*Philia*) – a force of attraction and combination; and Strife (*Neikos*) – a force of repulsion and separation.

Taking these two fundamental forces that from the basis of his insights for the cosmos; he developed his own terminology for the Cosmic Cycle which, briefly, is about the eternal circulation of both these abstracts, about which he wrote, "A twofold tale I shall tell: at one time it grew to be one only from many, and at another again it divided to be many from one. There is a double birth of what is mortal, and a double passing away; for the uniting of all things brings one generation into being and destroys it, and the other is reared and scattered as they are again being divided. And these things never cease their continuous exchange of position, at one time all coming together into one through Love, at another again being borne away from each other by Strife's repulsion".

In terms of our current topic i.e. Light; Empedocles was notably the first who raised the question about the nature and respective aspects about Light. Imagination or thinking of Empedocles understood Light as something which has identical properties to both Sun and Eyes. His ideology, in some sense says that the Light of Sun first strikes or arrives amidst space and then to our Eyes. This idea seems to be obviously rational. Although, there was a time when Empedocles himself doubted or misinterpreted his postulation because earlier he thought that eyes have this luminous property similar to sun, but if this is true then we can see in night also, which is absolutely not true and then he came to a very close ideology on Light which matches with the reality that we can't see in night.

His insight or philosophy over Light is obviously not fully considerable for modern sciences in the way that it could contribute in quantum physics; nevertheless his thinking allowed the legends of forthcoming generations including Gorgias, Socrates, Plato, Aristotle, Euclid, Lucretius, Johannes Kelper, Isaac Newton, Michael Faraday, James Clerk Maxwell, etc., to investigate the concept of "Light" further for the betterment of sciences and humanity, which led to the foundation of a lot of modern theories like quantum physics, Theory of Relativity, electromagnetism, etc.

7
Aristotle

The law is reason, free from passion. - Aristotle

Aristotle (384 BC-322 BC) was a Greek philosopher and one of the greatest polymaths of all time. He is sometimes also regarded as the father of certain realms of sciences in addition to being regarded as the father of modern philosophy. He was a student of Plato and the mentor to Alexander the Great. His contributions were in various different fields including philosophy, physics, logic, physiology, politics, economics, biology, etc. He has his own legacies in different areas including Aristotelian physics, Aristotelian philosophy and his treatise as Theory of the Soul, etc. All of these played a vital role in promoting sciences at an entirely different and higher level of thinking for modern ages. For example, in the account of his famous Aristotelian Physics, there was a separate concept of Gravity which roughly says that all bodies move towards their natural places or in other words, all objects have a kind of propensity to fall towards the center of the Universe, and also postulated that if an object fell at any peculiar constant speed, lighter ones reach earlier on being released whereas the heavier ones fall swiftly in proportion to their mass.

Although some of his conclusions were not proved entirely correct, his legacies made many other eminent contemplate over his work. For instance, in the case of gravity, Galileo Galilei later corrected his interpretation and made it more profound. Similarly, there are plenty of works on the attribution of Aristotle, which inspired scientists to explore the many mysteries of the Universe.

Specifically on Light, he believed that it is some sort of medium as the disturbance in one of the "four elements" from Empedocles i.e. 'air'. He was also influenced from the point of views of Empedocles and hence, in one of his legendary books, i.e. "De Sensu", Aristotle defined Empedocles' perception for Light as: "Empedocles says that the light from the Sun arrives first in the intervening space before it comes to the eye, or reaches the Earth. This might plausibly seem to be the case. For whatever is moved through space, is moved from one place to another; hence, there must be a corresponding interval of time also in which it is moved from the one place to the other. But any given time is divisible into parts; so that we should assume a time when the sun's ray was not as yet seen, but was still travelling in the middle space".

8
Euclid

The laws of nature are but the mathematical thoughts of God. - Euclid

Euclid was a Greek mathematician who intensively contributed in the field of geometry, algebra and algorithms. The kind of theories developed by him were quite ahead of his time, he defined certain very fundamental and original ideas which are now applied in various branches of mathematics like number theory, group theory, algebraic geometry etc. He is also regarded as the 'Father of Geometry', which he surely is. He was one of the most influential mathematicians ever in the history, his contributions also lead to certain very foundational applications into other fields like physics, fundamental computational aspects, etc. In fact, his ideas even contributed to quantum physics in a very profound way, with a complete theory abbreviated as Euclidean Field Theory or Statistical Field Theory; or in the case of Polymers known as the Polymer Field Theory, which determines the statistical or analytical characteristics of polymer in different charge conditions. One of the other bewildering concepts that arise from the intuition of this legendary man were the consequence of Euclidean Space, which in turn served as foundation to various modern theories in theoretical physics like Minkowski Space, Einstein's Theory of Relativity, etc.

As the motive of writing this book is specially towards the benefit of the general public, hence an additional note (just for the sake of foundational knowledge), it should be known that this concept of Euclidean Space is extracted from the geometrical interpretations of Euclid, which strictly satisfies under the domain of following axioms and postulates respectively:

Axioms:

- 1st Axiom: Things which are equal to the same thing, are also equal to one another.
- 2nd Axiom: If equals are added to equals, the whole are equal.
- 3rd Axiom: If equals be subtracted from equals, the remainders are equal.
- 4th Axiom: Things which coincide with one another are equal to one another.
- 5th Axiom: The whole is greater than the part.

Postulates:

- 1st Postulate: To draw a line from any point to any point.
- 2nd Postulate: To produce a finite straight line continuously in a straight line.
- 3rd Postulate: To describe a circle with any center and distance.
- 4th Postulate: That all right angles are equal to one another.
- 5th Postulate: That, if a straight line falling on two straight lines make interior angles on the same side less than two right angles, the two straight lines, if produced indefinitely, meet on that side of which are the angles less than the two right angles.

But another great and well-known German mathematician, David Hilbert (1862-1943), in his 1899 book named "Grundlagen der Geometrie", defined some more generalized axioms as a set of 20 assumptions, which are more rigorous than that of Euclid; and as per the "Foundations of Geometry (Grundlagen der Geometrie)" these axioms are classified as follows:

- Axioms of Incidence:

1. For every two points A, B there exists a line "a" that contains each of the points A, B.
2. For every two points A, B there exists no more than one line that contains each of the points A, B.
3. There exist at least two points on a line. There exist at least three points that do not lie on a line.

4. For any three points A, B, C that does not lie on the same line there exist a plane "α" that contains each of the points A, B, C. For every plane there exists a point which it contains.
5. For any three points A, B, C that do not lie on one and the same line there exists no more than one plane that contains each of the three points A, B, C.
6. If two points A, B of a line a lie in a plane α, then every point of a lies in the plane α.
7. If two planes α,β have a point A in common, then they have at least one more point B in common.
8. There exist at least four points which do not lie in a plane.

- Axioms of Order:

1. If a point B lies between a point A and a point C, then the points A, B, C are three distinct points of a line, and B then also lies between C and A.
2. For two points A and C, there always exists at least one point B on the line AC such that C lies between A and B.
3. Of any three points on a line, there exists no more than one that lies between the other two.
4. Let A, B, C be three points that do not lie on a line and let "a" be a line in the plane ABC which does not meet any of the points A, B, C. If the line apasses through a point of the segment AB, it also passes through a point of the segment AC, or through a point of the segment BC.

- Axioms of Congruence:

1. If A, B are two points on a line a, and A' is a point on the same or on another line a' then it is always possible to find a point B' on a given side of the line a' through A' such that the segment AB is congruent or equal to the segment A'B'. In symbols AB = A'B'.
2. If a segment A'B' and a segment A"B", are congruent to the same segment AB, then the segment A'B' is also congruent to the segment A"B", or briefly, if two segments are congruent to a third one they are congruent to each other.
3. On the line a, let AB and BC be two segments which except for B have no point in common. Furthermore, on the same or on another line a' let A'B' and B'C' be two segments which except for B' also have no point in common. In the case, if AB = A'B' and BC = B'C' then AC = A'C'.
4. Let$\angle$ (h,k) be an angle in a plane α and a' is a line in a plane α' and let a definite side of a' in α' be given. Let h' be a ray on the line a' that emanates from the point O'. Then there exists in the plane, one and only one ray k' such that the $\angle$ (h,k) is congruent or equal to the $\angle$ (h',k')& simultaneously, all interior point of the $\angle$ (h',k') lie on the given side of a'. Symbolically $\angle$ (h,k) = $\angle$ (h',k'). Every angle is congruent to itself, i.e., $\angle$ (h,k) = $\angle$ (h,k) is always true.
5. If, for two triangles i.e. triangle (ABC) and triangle (A'B'C') the congruence $\angle$ ABC = $\angle$ A'B'C' is also satisfied.

- Axiom of Parallels:

1. (Euclid's Axiom) Let a be any line and A, a point not on it. Then there is at most one line in the plane, determined by "a" and "A", that passes through "A" and does not intersect "a".

- Axioms of Continuity:

1. (Archimedes' Axiom or Axiom of Measure) If AB and CD are any segments, then there exists a number n such that n segments CD constructed contiguously from A, along the ray from A through B, will pass beyond the point B.
2. (Axiom of Line Completeness) An extension of a set of points on a line with its order and congruence relations that would preserve the relations existing among the original elements as well as the fundamental properties of line order and congruence that follow from first three axioms and from a just previous one; is entirely impossible.

Now, as far as the Theory of Light is concerned, the thoughts of Euclid were expressed in a mathematical way which he used to deal as the systems in Optics. In this realm, he studied the behaviour and properties of Light with mathematical essence, in which he postulated that Light travels in a straight line. He was also sceptical about some existing and previous philosophies which explore vision or sight as a kind of beam from the eye. Long after Euclid, in relation to this philosophy, after Empedocles and Aristotle, there was another great Roman philosopher and poet named Titus Lucretius Carus who was highly influenced by the philosophies or thoughts of both Empedocles and Aristotle, and wrote in his book "De Rerum Natura",

"The light and heat of the Sun, both are composed of infinitesimal atoms, which, when they are shoved off, lose no time in shooting right across the inter-space of air in the direction imparted by propel".

The work of Euclid was fundamental with a mathematical beauty which influenced a lot of brilliant people that came after him and thus his contribution in the establishment of the Light-Theory is very significant from both theoretical and mathematical perspectives which not only contributed to the study of Optics and beyond in physics, but also to quantum mechanics.

9
René Descartes

The greatest minds are capable of the greatest vices as well as of the greatest virtues. - René Descartes

René Descartes (1596-1650) was a French philosopher, mathematician and physicist who had some extraordinary and far-sighted capabilities. His astounding expertise ruled and guided the scientific world for a long time and some of his theories and philosophies are still quite applicable and admired in the fields. Due to his phenomenal services to the subjects, he is regarded as one of the founders of modern philosophy. His legacies and aura have been a kind of paradigm to the next generations that were fascinated towards sciences. Some of his most notable ideas were analytic geometry, Cartesian coordinate system, causal adequacy principle, mind-body problem, the wax argument, etc. In fact, some scientists use his mind-body problem and causal adequacy principle into the unification of quantum mechanics and Consciousness, thus his principles in some sense also build a foundation for the modern interpretations in the quantum realm. The mind-body problem fundamentally deals with the abstract relationship between the Consciousness and thoughts in the human brain. This concept is sometimes also abbreviated as Cartesian dualism. The next one, the causal adequacy principle, briefly says that anyway, an object must accommodate such least degrees of reality as intrinsically contained by the object itself.

Both of these sometimes related with the 'Observer Effect' in quantum mechanics (concerns with the atomic or sub-atomic particles) which says that a meagre observation certainly changes the observed reality. This 'observer effect' is also linked with some cutting-edge foundational principles of quantum mechanics such as Heisenberg's Uncertainty Principle; apart from it, this also has very significant applications in particle physics too. Hence, the importance of Descartes' insights are quite useful, even in the modern times including mathematics, neuroscience, physics etc.

Coming to his postulations over light, they are already defined in one of the previous parts, as originating in pre-Socratic times relative to the distinctive reality. But now a few more significant points are going to be added into the previous one to make the concept clearer. Aristotle was one of those great people by whom René Descartes was highly influenced. Hence, he was favouring Aristotle's view-point over Light. He hypothesizes an invisible fabric or substance, which he termed Plenum, percolating the universe. Like Aristotle, he postulated that Light is a disturbance which travels through the Plenum, similar to the way a wave travels by the means of water. This was a kind of raw explanation by him for Light, but he filled the gap in one of his books which had revolutionary insight on Optics.

The book is named as "La dioptrique" and it is that piece of his legacy which is concerned with the origins of "Laws of Optics" lead by Sir Isaac Newton later. René Descartes mainly offered an explanation for the phenomena of Reflection and Refraction, which at the end turned out in mathematical and geometrical interpretations. His model used a tennis ball to observe its behaviour when hit towards a surface by a racquet and this became the key to explore the laws of Reflection and Refraction as the set of formation of models. Eventually, he got to three different conclusions which are:

- The decisive factor of both motion and direction of motion (of ball) are comprehended.
- The ball's motion can be identified in both horizontal and vertical components.
- The motion of ball can be shown with geometrical interpretations as straight lines.

This was just for additional knowledge, but the key point is to make you familiar with the significance of René Descartes' contributions towards the nature of light and beyond, which played a vital role in the evolution of quantum mechanics in the modern metaphysical aspects and other interpretations, especially the birth of the conceptualization of Light – it being the most fundamental point with which the evolution of quantum physics began.

10
Isaac Newton

Truth is ever to be found in simplicity, and not in the multiplicity and confusion of things. - Isaac Newton

Isaac Newton (1643-1727) is certainly one of the greatest scientists that ever walked on earth. He worked in plenty of different areas of sciences and made notable contributions in the fields of mathematics, physics, theology, astronomy etc. He also widely credited to have discovered Calculus, independently to Gottfried Wilhelm Leibniz. His discoveries have influenced the scientific world enormously right up to the present age. The legacies of Newton have influenced some of the most brilliant and gifted scientists of all times, like Joseph Louis Lagrange, James Clerk Maxwell, Albert Einstein etc. Louis Lagrange considered him the greatest genius who ever lived, the great English poet Alexander Pope wrote for Newton,

Nature and nature's laws lay hid in night;

God said "Let Newton be" and all was light.

Newton's contribution to the theory of Light is quite revolutionary in both the sense of Optics or Light in itself as a singular entity. Ole Rømer (1644–1710) was an astronomer who came up with a demonstration (Rømer's determination of the speed of light) in the year 1676 and favoured the concept of finite speed of light. It was estimated that the Light takes approximately 22 minutes to travel a distance equal to that of the diameter of Earth's orbit around the sun.

His theory got very few supporters, but Isaac Newton was one of the foremost among them, who then took his idea to propose the estimate time taken by Light to travel from Sun to Earth as between 7-8 minutes. Newton's further treatment of the demonstration of Rømer can be found in his book "Opticks", written in the year 1704.

His other contributions are in the operational applications of Light into the form of reflection, refraction or other kind of aspects in the realm of Optics. I do not find it suitable to define them one by one separately as our current aim is to explore the evolution of Quantum physics. But for some foundational knowledge, a part of Isaac Newton's own letter is presented below, which he had sent to the publisher from Cambridge in order to be communicated to the Royal Society for his new ideas over Light, colors and beyond. The whole letter [attributed to the Philosophical

Transactions of the Royal Society, No. 80 (19 Feb. 1671/2), pp. 3075-3087]is very long but as it's just for the sake of superficial knowledge (or imagination), let's look at a brief part of the letter –

[from pp. 3081-3085 among all pages in the Philosophical Transactions of the Royal Society, No. 80 (19 Feb. 1671/2) i.e. pp. 3075-3087] in which there are certain peculiar postulates regarding his insights over the other relative of Light (A few words are now much evolved & corrected as compared to that time of Sir Isaac Newton (like appeareth is as appeared, prisme as prism etc.; but I'm not changing anything from the kind of Language described by Newton at that time) :

> "1. As the Rays of light differ in degrees of Refrangibility, so they also differ in their disposition to exhibit this or that particular colour. Colours are not *Qualifications of Light,* derived from Refractions, or Reflections of natural Bodies (as 'tis generally believed,) but *Original* and *connate properties,* which in divers Rays are divers. Some Rays are disposed to exhibit a red colour and no other; some a yellow and no other, some a green and no other, and so of the rest. Nor are there only Rays proper and particular to the more eminent colours, but even to all their intermediate gradations.
>
> 2. To the same degree of Refrangibility ever belongs the same colour, and to the same colour ever belongs the same degree of Refrangibility. The *least refrangible* Rays are all disposed to exhibit a *Red* colour, and contrarily those Rays, which are disposed to exhibit a *Red* colour, are all the least refrangible: So the *most refrangible* Rays are all disposed to exhibit a deep

Violet Colour, and contrarily those which are apt to exhibit such a violet colour, are all the most refrangible.

And so to all the intermediate colours in a continued series belong intermediate degrees of refrangibility. And this Analogy 'twixt colours, and refrangibility, is very precise and strict; the Rays always either exactly agreeing in both, or proportionally disagreeing in both.

3. The species of colour, and degree of Refrangibility proper to any particular sort of Rays, is not mutable by Refraction, nor by Reflection from natural bodies, nor by any other cause, that I could yet observe. When any one sort of Rays hath been well parted from those of other kinds, it hath afterwards obstinately retained its colour, notwithstanding my utmost endeavours to change it. I have refracted it with Prismes, and reflected it with Bodies, which in Day-light were of other colours; I have intercepted it with the coloured film of Air interceding two compressed plates of glass; transmitted it through coloured Mediums, and through Mediums irradiated with other sorts of Rays, and diversly terminated it; and yet could never produce any new colour out of it. It would by contracting or dilating become more brisk, or faint, and by the loss of many Rays, in some cases very obscure and dark; but I could never see it changed *in specie.*

Yet seeming transmutations of Colours may be made, where there is any mixture of divers sorts of Rays. For in such mixtures, the component colours appear not, but, by their mutual allaying each other, constitute a middling colour. And therefore, if by refraction, or any other of the aforesaid causes, the difform Rays, latent in such a mixture, be separated, there shall emerge colours different from the colour of the composition. Which colours are not New generated, but only made Apparent by being parted; for if they be again intirely mix't and blended together, they will again compose that colour, which they did before separation. And for the same reason, Transmutations made by the convening of divers colours are not real; for when the difform Rays are again severed, they will exhibit the very same colours, which they did before they entered the composition; as you see, *Blew* and *Yellow* powders, when finely mixed, appear to the naked eye *Green,* and yet the colours of the Component corpuscles are not thereby really transmuted, but only blended. For, when viewed with a good Microscope, they still appear *Blew* and *Yellow* interspersedly.

5. There are therefore two sorts of Colours. The one original and simple, the other compounded of these. The Original or primary colours are, *Red, Yellow, Green, Blew,* and a *Violet-purple,* together with Orange, Indico, and an indefinite variety of Intermediate gradations.

6. The same colours in *Specie* with these Primary ones may be also produced by composition: For, a mixture of *Yellow* and *Blew* makes *Green*; of *Red* and *Yellow* makes *Orange*; of *Orange* and *Yellowish green* makes *yellow*. And in general, if any two Colours be mixed, which in the series of those, generated by the Prisme, are not too far distant one from another, they by their

mutual alloy compound that colour, which in the said series appeareth in the mid-way between them. But those, which are situated at too great a distance, do not so. *Orange* and *Indico* produce not the intermediate Green, nor Scarlet and Green the intermediate yellow.

7. But the most surprising, and wonderful composition was that of *Whiteness*. There is no one sort of Rays which alone can exhibit this. 'Tis ever compounded, and to its composition are requisite all the aforesaid primary Colours, mixed in a due proportion. I have often with Admiration beheld, that all the Colours of the Prisme being made to converge, and thereby to be again mixed as they were in the light before it was Incident upon the Prisme, reproduced light, intirely and perfectly white, and not at all sensibly differing from a *direct* Light of the Sun, unless when the glasses, I used, were not sufficiently clear; for then they would a little incline it to *their* colour.

8. Hence therefore it comes to pass, that *Whiteness* is the usual colour of *Light;* for, Light is a confused aggregate of Rays indued with all sorts of Colors, as they are promiscuously darted from the various parts of luminous bodies. And of such a confused aggregate, as I said, is generated Whiteness, if there be a due proportion of the Ingredients; but if any one predominate, the Light must incline to that colour; as it happens in the Blew flame of Brimstone; the yellow flame of a Candle; and the various colours of the Fixed stars.

9. These things considered, the *manner*, how colours are produced by the Prisme, is evident. For, of the Rays, constituting the incident light, since those which differ in Colour proportionally differ in Refrangibility, they by their unequall refractions must be severed and dispersed into an oblong form in an orderly succession from the least refracted Scarlet to the most refracted Violet. And for the same reason it is, that objects, when looked upon through a Prisme, appear coloured. For, the difform Rays, by their unequal Refractions, are made to diverge towards several parts of the *Retina*, and there express the Images of things coloured, as in the former case they did the Suns Image upon a wall. And by this inequality of refractions they become not only coloured, but also very confused and indistinct

10. Why the Colours of the *Rainbow* appear in falling drops of Rain, is also from hence evident. For, those drops, which refract the Rays, disposed to appear purple, in greatest quantity to the Spectators eye, refract the Rays of other sorts so much less, as to make them pass beside it; and such are the drops on the inside of the Primary Bow, and on the outside of the Secondary or Exteriour one. So those drops, which refract in greatest plenty the Rays, apt to appear red, toward the Spectators eye, refract those of other sorts so much more, as to make them pass beside it; and such are the drops on the exteriour part of the *Primary*, and interiour part of the *Secondary* Bow.

11. The odd Phenomena of an infusion of *Lignum* Nephriticum, Leaf gold, Fragments of coloured glass,

and some other transparently coloured bodies, appearing in one position of one colour, and of another in another, are on these grounds no longer riddles. For, those are substances apt to reflect one sort of light and transmit another; as may be seen in a dark room, by illuminating them with similar or uncompounded light. For, then they appear of that colour only, with which they are illuminated, but yet in one position more vivid and luminous than in another, accordingly as they are disposed more or less to reflect or transmit the incident colour.

12. From hence also is manifest the reason of an unexpected Experiment, which Mr. Hook somewhere in his Micrography relates to have made with two wedg-like transparent vessels, fill'd the one with red, the other with a blew liquor: namely that though they were severally transparent enough, yet both together became opake; For, if one transmitted only red, and the other only blew, no rays could pass through both.

13. I might add more instances of this nature, but I shall conclude with this general one, that the Colours of all natural Bodies have no other origin than this, that they are variously qualified to reflect one sort of light in greater plenty then another. And this I have experimented in a dark Room by illuminating those bodies with uncompounded light of divers colours. For by that means any body may be made to appear of any colour. They have there no appropriate colour, but ever appear of the colour of the light cast upon them, but yet with this difference, that they are most brisk and vivid in the light of their own day-

> light-colour. *Minium* appeareth there of any colour indifferently, with which 'tis illustrated, but yet most luminous in red, and so *Bise* appeareth indifferently of any colour with which 'tis illustrated, but yet most luminous in blew. And therefore *Minium* reflecteth Rays of any colour, but most copiously those indued with red; and consequently when illustrated with day-light, that is, with all sorts of Rays promiscuously blended, those qualified with red shall abound most in the reflected light, and by their prevalence cause it to appear of that colour. And for the same reason *Bise*, reflecting blew most copiously, shall appear blew by the excess of those Rays in its reflected light; and the like of other bodies. And that this is the intire and adequate cause of their colours, is manifest, because they have no power to change or alter the colours of any sort of Rays incident apart, but put on all colours indifferently, with which they are inlightned".

This was it; from Newton's own writings. His work on Light and beyond certainly revolutionized the field of Physics and understanding of nature forever.

11
Christiaan Huygens

I believe that we do not know anything for certain, but everything probably. – Christiaan Huygens

Christiaan Huygens (1629-1695) was a Dutch physicist who is best known for his outstanding contributions in various fields of study like physics, mathematics, and astronomy, etc. Like Isaac Newton, he was also one of the great scientific figures of his time. Due to his astonishing ideas, he is regarded as the founding father of mathematical physics which plays a great role in quantum physics as well. In physics, some of his ideas are also applicable in the regime of Newtonian physics. As it's now known that quantum mechanics is the probabilistic way to determine the things at a very diminutive level, there is a peculiar contribution of Christiaan Huygens, which was a kind of early treatise on the probability theory, such that along with Blaise Pascal and Pierre de Fermat he too became one of the pioneers of probability theory.

Coming to his contribution on the theory of Light, it is unequivocally enormous. After the foundational role played by the astronomer Ole Rømer and later Isaac Newton, Huygens came with his insights on the Rømer's determination of the speed of light, already described briefly previously in consideration of the case of Isaac Newton. Huygens found out a method to attain the speed of light with certain precision, he basically somehow amalgamates Rømer's way of determination with another estimation regarding the diameter of the orbit of Earth and thus he calculates the speed of light as approximately 26% less then what the current value is. But anyway, it was a great attempt which opened the pathway to the establishment of certain new ideas and their further development.

This was about his contribution to the Theory of Light. His other contributions are just as fantastically extraordinary which still rule the scientific world in many ways and lead to the evolution of physics in the modern age. One of his major contributions is the Wave Theory of Light. It has already been briefly discussed in some of the initial chapters. Apart from Huygens, Robert Hooke was also made some legendary contributions to the wave theory of Light. But Huygens was the scientist who deeply investigated the properties of waves attributed to Light and beyond applications.

Huygens in his "Treatise on Light" written in the year 1690 hypothesized a universal medium for the propagation of light which is called the "Luminiferous aether". This sparked discussion among top order scientists and experimentalists who even came to accept his hypothesis, but it was later discarded by Albert Einstein his Theory of Special Relativity. Lincoln Barnett, in his book "Universe and Dr. Einstein" writes, "Among those who pondered the enigma of the Michelson—Morley experiment was a young patent office examiner in Berne, named Albert Einstein. In 1905, when he was just twenty-six years old, he published a short paper suggesting an answer to the riddle in terms that opened up a new world of physical thought. He began by rejecting the ether theory and with it the whole idea of space as a fixed system or framework, absolutely at rest, within which it is possible to distinguish absolute from relative motion. The one indisputable fact established by the Michelson—Morley experiment was that the velocity of light in unaffected by the motion of the earth. Einstein seized on this as a revelation of universal law. If the velocity of light is constant regardless of the earth's motion, he reasoned, it must be constant regardless of motion of any sun, moon, star, meteor, or other system moving anywhere in universe. From this he drew a broader generalization, and asserted that the laws of nature are the same for alls uniformly moving system. This simple statement is the essence of Einstein's Theory of Special Relativity. It incorporates the Galilean Relativity Principle, which states that mechanical laws are the same for all uniformly moving systems. But its phrasing is more comprehensive; for Einstein was thinking not only of mechanical laws but of the laws governing light and other electromagnetic phenomena. So he lumped them together in one fundamental postulate: all the phenomena of nature, all the law of nature, are the same for all systems that move uniformly relative to one another".

But later, when he delivered a lecture at the University of Leiden in the year 1920, he said –

"Recapitulating, we may say that according to the general theory of relativity space is endowed with physical qualities; in this sense, therefore, there exists aether. According to the general theory of relativity space without ether is unthinkable; for in such space there not only would be no propagation of light, but also no possibility of existence for standards of space and time".

This statement makes a lot of sense, one that he is accepting the aether; but the other one is – perhaps he had some other kind of properties or prerequisites of aether (or ether) in his mind similar to quantum physics.

Well, this was just a fascinating history but now the principle of Huygens, widely known as the "Huygens–Fresnel Principle". Although, there is absolutely no relationship between the two scientists the principle is named after; in fact both scientists belong to different times and different countries. But Fresnel's work extends Huygens' theory in a better version with more applications, specifically towards the rectilinear propagation of light and corresponding diffraction effects. Huygens' theory was basically capable of providing the theoretical and mathematical explanations on some fundamental aspects like linear and spherical propagations in unison with the laws of reflection and refraction respectively.

There is a very long history regarding the origin of Huygens' perceptions, which may be irrelevant to quantum theory, but there is one point where the Huygens–Fresnel

Principle shares a relationship with the modern age's quantum field theory. But before having a brief understanding of this, one needs to be familiar with at least two of the most significant postulates of the principle, given below:

- Every distinctive point on the wavefront (a surface over which an optical wave has a constant phase, i.e. an angle representing the number of periods attributed to the respective periodic function) is a new source of disturbance known as secondary wavelets. This travels in arbitrary directions under the domain of medium with the speed of original wave (or somewhat velocity of light).
- At any span, the packet of secondary wavelets gives rise to a new wavefront at that peculiar point.

The principle contributes to quantum field theory through its working as one of the aftermaths of space homogeneity. Though the pure concept of homogeneity is entirely linked with the topological aspects which are quite tough to explain in simple language, but as far as the analogy of Huygens-Fresnel Principle with quantum field theory is concerned, Space homogeneity (or Space-space homogeneity) is one of the fundamentals of quantum field theory, which suggests the wave-function as superposition of all waves propagating in all existing spatial paths. Apart from this, the principle has more applications in quantum field theory, which act under certain local transformations mathematically. French mathematician Jacques Salomon Hadamard is a notable personality who was also the first to decode that the Huygens-Fresnel Principle holds breakdown when the number of spatial dimensions are even, from which observation and result he continued to define the analogies in the form of a fundamental mathematical framework, which became one of the biggest reason for the direct correlation of the principle with Quantum world. But entire mathematical explanation will not be a suitable to be illustrated here for the general public as only a few people understand it.

Moving further, other than Augustin-Jean Fresnel, Thomas Young was another great scientist who worked for the enhancement of Huygens' theory. One can now see how the legacy of Christiaan Huygens came to be a masterpiece in the realm of theoretical physics and particularly quantum physics; his great contribution towards the advancement of quantum field theory has now developed in such a way that it has driven a lot of significant applications both theoretically and experimentally.

12
Michael Faraday

Nothing is too wonderful to be true if it be consistent with the laws of nature. - Michael Faraday

Michael Faraday (1791-1867) was an English physicist and experimental scientist who made some of the greatest scientific contributions for the betterment of humanity and sciences. His contributions also played a significant role in physical chemistry; in fact he is one of the pioneers of electrochemistry. He was also one of those great scientists who had a big hand in developing electromagnetism at a higher cognition, for e.g. he discovered electromagnetic induction. The most inspirational fact about Faraday is that he never had a formal education at the higher level, it is said that he had formal education only up to the age of 13-14. It is phrased in history of sciences that the greatest mind in human history, Albert Einstein only had three pictures in his study which were of Isaac Newton, James Clerk Maxwell and Michael Faraday. Like Einstein (as previously mentioned), Faraday also raised a question against the concept of Luminiferous aether, which was taken into account by both Christiaan Huygens and then Augustin Jean Fresnel in the wave theory of Light. Faraday suggested that perhaps Light is an upshot of the shivering of physical lines of force rather than any sort of latent aspect referring to "aether".

But anyway, he was in favour of the fact considering certain kind of vibrations for the existence of Light, combined with any distinct disturbance in its propagation.

In the year 1845, Faraday had an insight to relate magnetic field and polarized light, i.e. the wave with peculiar geometrical orientation of oscillations. This relationship he built from his experimental observation which was about the quivering plane is incident as the linearly polarized light beam on a glass-piece, which holds spin when a magnetic field was imposed in the direction of propagating beam. This experiment or observation is known as Faraday Effect or Magneto-optic Faraday Effect. This has the application of one of the most significant experimental aspect lead by Faraday himself, i.e. the Faraday Rotator, which is attributed to the Faraday effect itself; both the core effect and its related rotator played many notable roles in various fields with the passage of time for instance in the single directed transmission of light, i.e. Optical diode with the help of rotator's application as the amplitude modulation of light, etc. His work was the first ever manifestation of the insight that Light has some relation with the phenomena of electromagnetism.

As quantum physics has this abstract touch of superposition principle in its foundational perceptions, we can relate the Faraday Effect with this concept too because the Faraday Rotator when given rise to its role in the experiment of core effect, the polarized beam which is fundamentally annular at that particular point of time, has a concomitant of superposition and the proper term i.e. circularly polarized light beam has a very significant role in the electrodynamics and thus this side of Faraday's interpretation not only give its contribution to the field of quantum mechanics on a general basis but also in one of the advanced version of quantum field theory, specifically in the domain of Particle physics, i.e. Quantum Electrodynamics. Apart from this, it also plays very significant roles in the area of telecommunication and beyond on the basis of its consecutive progress in the exclusive realms of quantum physics. And so the legacies of Michael Faraday possess the intense capability to bring about bewildering scientific and technological aspects in the future and perhaps play a consequential role in quantum computing too.

13
James Clerk Maxwell

It is a good thing to have two ways of looking at a subject, and to admit that there are two ways of looking at it. – James Clerk Maxwell

James Clerk Maxwell (1831-1879) was a Scottish physicist and mathematician who elegantly contributed in the several fields and disciplines of study. He is generally known for his outstanding research on the works of Michael Faraday. His study has furthered a lot of works and conclusions by Faraday into advanced regimes of study with a mathematical touch. He was a great man who brought up the result that electricity, magnetism and light are nothing but different expression of identical phenomena, like Albert Einstein who postulated the truth of space and time as different manifestations of the same thing. After Isaac Newton's systematic conceptualization Gravity, Maxwell's unification is the most highly regarded one in the history of physics. His work became one of the most significant and cogent foundational aspect in the development of Einstein's Theory of Special Relativity. Due to his services to the field, he is regarded as the father of modern classification of light which is the classical theory of electromagnetism, and modern electrical engineering. It is said that Einstein greatly regarded Maxwell's legacy as the most profound description in the account of physics after Isaac Newton. There is a popular incident that when Einstein visited Cambridge in 1922, the entertainer told him that he came up with another revolution in sciences as he stood on the shoulders of Newton; Einstein replied by saying – "No, I don't. I stand on the shoulders of Maxwell"; this is the kind of respect that Maxwell is accorded with in the world of science. It can be said that if Maxwell was ever asked the same question he would've surely said the same for Michael Faraday. In fact, Faraday's ideology about field lines wasn't fully accepted until the mathematical & theoretical findings of Maxwell became a part of it.

Faraday's insights about the vibration of a ray or beam, was also not accepted until it got the support of Maxwell's explanations and formulations. Most profoundly, in 1865; James Clerk Maxwell investigated Faraday's experimental observations with respect to light in his own theoretical & mathematical ways which came out as the "The Dynamical Electromagnetic Theory of Light", which had a compilation of several vital conclusions that played a great role in the establishment of the theory of light. For example, a seminal description of electromagnetic theory of light, the travelling of electric and magnetic fields as waves with the velocity of light as a medium of space etc. There were a lot of experimental aspects also that came out of it, for example, the Heinrich Rudolf Hertz's laboratory confirmation of Faraday's thought on ray vibrations in the year 1888 which were found to be as mathematically profound as defined by J C Maxwell.

Maxwell always regarded the legacies of Faraday so highly that he even described his name and relative thoughts in that paper certain times as,

"The conception of the propagation of transverse magnetic disturbances to the exclusion of normal ones is distinctly set forth by Professor Faraday in his 'Thoughts on Ray Vibrations'. The electromagnetic theory of light, as proposed by him [Faraday], is the same in substance as that which I have begun to develop in this paper, except that in 1846 there were no data to calculate the velocity of propagation".

And,

"Faraday discovered that when a plane polarized ray transverses a transparent diamagnetic medium in the direction of the lines of magnetic force produced by magnets or currents in the neighbourhood, the plane of polarization is caused to rotate."

Apart from Maxwell's contribution in the modernization of Light, there are plenty of his contributions which lead to the development of the quantum mechanics, specifically in favour of corresponding probability distributions, statistical mechanics and thermodynamically established systems etc. All of these have been fundamentally taken into consideration in over the next few chapters, when unfolding the patterns of statistical mechanics to be specific.

Just like the special relativity, Maxwell's several different works have been proven highly cardinal for not only mathematics like in tensor analysis, but also for the qualitative enlargement of quantum mechanics, which will always be highly significant for more futuristic developments via theoretical physics.

14
Albert Einstein

Coincidence is God's way of remaining anonymous.
- Albert Einstein

Albert Einstein (1879-1955) was the ultimate physicist, mathematician, philosopher and much more; whose name today has become synonymous with genius. His contributions to subjects are milestones in the world of theoretical physics and mathematics. The persona belongs to the class of geniuses like Aristotle, Galileo Galilei, Christiaan Huygens, Isaac Newton, Michael Faraday and James Clerk Maxwell. Even the most brilliant scientists of his time were awestruck by his Genius, his scientific theories and philosophical ideologies pushes humanity to this day and will keep on enlightening the mankind. One cannot help but wonder whether they are the humanly creation.

As far as the conceptualization of Light is concerned he played a great role in its development as a theory with his famous photoelectric effect. More related aspects of the photoelectric effect are discussed in the Einstein-centric chapters constructed around his enormously magnificent discoveries and revelations. But as a brief note apart from classical theory regarding photoelectric effect, which was about the transfer of energy from light to an electron, Einstein's perception (also attributed to Max Planck) over the photoelectric effect is slightly different which takes effect as the emission of charge carriers or electron when the action of any electromagnetic radiation (like Light) imposed on the specific material (like surface of any metal sheet). When electrons are emitted in this way they are called photoelectrons. Einstein was also indicating towards the classical view as the "Corpuscular Theory of Light", which argued in a similar way as him - the light beam as not a wave but a form of discrete wave packets, because the continuity of a wave won't allow the phenomena hold by the photoelectron which is the most fundamental side of this effect. From this theory, the concept of photon also came into the consideration as those discrete wave packets are nothing but photons and the flow of these forms a wave.

Part 3:
Dealing with the Quantum Reality

15
Vacillation in the Theory

The best that most of us can hope to achieve in physics is simply to misunderstand at a deeper level.
- Wolfgang Pauli

It has already been defined plenty of times that the account of quantum physics is full of superposition and possibilities in some sense. Or in simple language it can be understood in the way that if on one side, Isaac Newton were to define exactly what the place is going to be of any object after certain period of time with respect to any peculiar frame when all its previous physical system of records like momentum, deviation of density, etc. have been taken into consideration, then on the other hand, Werner Heisenberg or Erwin Schrödinger will only tell the possibilities of the several places of the same object after that same definite time span. Although it's not a strange thing, this is just the most fundamental underlying rule of the game known as "Quantum Mechanics". The foundations of quantum theory have the origin of probabilistic behaviours and the great scientists like Albert Einstein when they have tried to make it deterministic their insights have too later been converted into the same way as deep-rooted facts of the theory is at this current time for example, the wave-particle duality. It's definitely not the case that quantum mechanics is a failure of science; it's just that the current native properties of quantum theories don't possess the capabilities to determine firm reality because it violates both determinism and realism on a very elementary level. By Determinism it is not meant that something is full of superstitions or any relevant supernatural thing or something previously decided by any superior-entity; but refers to that kind of compilation of laws, theories, and formulations etc. which have very definable construction. On the other side, Realism here does not mean the artistic movement of France or anything like defining the reality that seems to be real; but entirely beyond from these superficial perceptions as here it is indicates towards the way of something universal regardless of the superposition.

The current fact is that quantum physics is strategically carved-up into two different realms, as attempts in two non-identical manners for the search of truth. One is physical that contains hypotheses, postulations, principles, theories, mathematical formulations, experimental aspects, etc.; and the other is metaphysical, which is constructed on the basis of ancient philosophies, certain effects on the basis of experiment or hypotheses, some direct conceptualizations, assumptions, etc. But the main catch is that both the realms are not taken in two different frames and a few people are just merely trying to combine both of them at any cost.

Although in both of these regimes, there are a few common notions like experiments with some precise interpretations, hypotheses and specified effects, etc. and thus, perhaps a few things can be correlated into that domain of rationality or suitable intuitiveness but not all for sure, and this thing is not understood by a lot of eminent people and hence they are trying their best to mix both these non-identical things due to the misconception of identical domain which is quantum mechanics. Actually the truth is there is no need to relate metaphysics with quantum explanations because metaphysics is already a part of study in the area of Philosophy, which is highly profound & rational too, at times.

A few intellectuals consider quantum physics a possible gateway to explain metaphysical aspects too; but before that they need to know what exactly metaphysics refers here because it has multiple answers that vary in opinions. It's not that they are wrong because they see consciousness too as a cause of superposition as the same thing is there in the natural behaviour of quantum mechanics too, therefore they relate them both. But one fact gets missed here and that fact is that consciousness is not any probabilistic game and hence metaphysics and physics must be considered separately; of course a few things can act seemingly in the identical range, but not actually exist.

The truth or reality is not any kind of set of consequences same as of any coin or a game board of probabilities etc. In all of these cases, the consequences are multiple but the origin of outcomes are not; like a coin is a single entity with two facets, but the frame of origin is identical, i.e. the coin itself. Similarly, there can be variegated facets to the reality but the entity remains with no chaos at initial point, whereas quantum mechanics holds its meaning only with the existence of chaos or randomness, although with certain patterns. With the help of quantum theory, we can happily get up to tending the reality but not the reality itself. Hence, the universal laws at any level can't be interlinked into a same domain of randomness or chaos held by quantum mechanics.

16
Close to Nature

Individuality is only possible if it unfolds from wholeness.
- David Bohm

Plenty of aspects have been taken in to consideration with the respect to quantum mechanics but it has still remained with certain amount of scope for modification for a higher level of accuracy. But, the thing is how far or close the quantum explanations are to the definition of the reality of nature. From previous chapters, it's already known that quantum physics is not facing any kind of failure in order to define the random behaviour including relative patters with it; but there has to be a definition or ideology of that level of perception that is fundamentally required to be attained which can make the field more profound as compared to now, thus explanations can have that development which make the quantum aspects clearer to define things or phenomena that plays a vital role in not only certain box of interpretations but also to elegance of theoretical and mathematical realms. For the sake of this, it is also necessary to exactly figure out that how much deviation quantum mechanics is showing in the regimes of its intrinsic elaborations and the truths which are already out there is theoretical physics, and this will result in helping the sciences to decode this mysterious universe in such a way that the formulations or theories can give rise to determinism within the domain as defined in one of the previous chapters.

There are a few things which indicate the distinction between current quantum elaborations as a limited range of explanation, for example:

- The Observer Effect :

  The observer effect is already briefly defined in one of the previous sections. This is a fundamentally rational attempt to explain the uncertainties lead by the formulations or interpretations of quantum mechanics. It is also linked with certain modern interpretations of quantum theory. But what if we see it with another perception, for instance let's first say that you have some kind of supernatural power to see the microscopic elements and you're currently reading this book with the help of a torch assuming electricity in your area, and after a few minutes you keep a bookmark on this page to read it later; now after some time, when you look for the book, it would be found in the same place where it was before.

> Now, assume that the book became half the size of what it was before, then also you'll find it in the same place. But what if we assume that the book becomes the size of an electron at the time when you already know the fact where the book's location is. Consider the fact that we are able to see things because of the reflection of light from that object which enters into our eyes and thus it forms a particular image on the retina. After a few minutes, you come to the same place to read the book and you shine the torch light at the place where the book was previously located, but this time you don't see the book because the emerged photon on the electron-sized book will lead an interaction which will deviate the path of that quantum-scaled book. When you check other sides, the same thing will keep taking the place.

Now is it any problem with sight or anything related to the consciousness or is it just a measurement effect? Well, that fact is crystal clear that it is only a measurement effect with rationality of phenomena. Although, it's tough to conceive the ideology that observer effect is only perhaps a kind of measurement effect in itself, which doesn't allow people to deal with quantum mechanics in the way that it should be. Apart from that, there is a saying on quantum physics by the famous American author Neale Donald Walsch, "Quantum physics tells us that nothing that is observed is unaffected by the observer. That statement, from science, holds an enormous and powerful insight. It means that everyone sees a different truth, because everyone is creating what they see". This saying is certainly making a thoughtful point about quantum mechanics and this has to be taken into the consideration as it may help the field take an extra step towards the reality's course of action attributed to the quantum world.

There is a serious possibility of any sort of discrepancy in the area of quantum mechanics regarding its measurement techniques, not every single time though as its mathematics holds the ability to be profound from various corners of technical elegance. The measurement effect can be seen in various interpretations like double-slit experiment too e.g. regarding the Heisenberg's uncertainty principle or beyond. But, the key point is that still quantum physics holds that capability to decode the deep nature with respect to its reality because the thing is – in plenty of uncertainties there always exists a single certainty which is responsible for the governance of the universe.

Albert Einstein still remains at the top with significant advises and suitable transforms in the field of quantum physics because the reason is we can't change the reality according to the existing theories. There is a possibility of another case regarding the experimental evidences in favour of existing hypotheses or principles in quantum world; and that possibility is that perhaps we have right experimental data or respective affirmations which are quite focused on the theoretical or mathematical aspects but the case may be with the interpretations regarding those theories or principles.

Part 4:
Initiatives for the New Explanation

17
Quantum Theory and The Black-body Revelation

Science cannot solve the ultimate mystery of nature. And that is because, in the last analysis, we ourselves are a part of the mystery that we are trying to solve. – Max Planck

Max Planck with his paper on black-body radiation in 1900, put forth a foundation of quantum physics in a way it had never been done before. Fundamentally, the black-body is a kind of idealized non-reflective physical body that absorbs all kind of incident electromagnetic radiation irrespective of any physical quantities like angle of incidence, frequency, etc., and the emission of the black-body within the domain of thermodynamically equilibrium state give rise to black-body radiation. Black-body radiation can also take place with the motion of particle inclusively, its thermal properties and thus, the black-body radiation also called as thermal radiation, which is also responsible for the conversion of body's internal energy into the electromagnetic energy.

This black-body radiation has a specific dependency in the form of continuous spectral energy distribution on the body's thermal state. The spectral energy distribution is a plot of energy v/s wavelength of light, which is also known as Frequency spectrum. Max Planck is neither the father of this concept of black-body Radiation nor the scientist who named it. The term 'black-body radiation' was first coined by the German Physicist Gustav Robert Kirchhoff. Kirchhoff is generally known for his contribution in electrical circuits but his more astonishing contributions are in the fields of Spectroscopy, phenomena of black-body radiation, thermodynamics and thermochemistry, etc. But the fact is not even Gustav Robert Kirchhoff was the first to think of the phenomena, though there is no question on the nomenclature for fathering and advancing the phenomena. But he was still the second person to identify it. The systematic development of the black-body Radiation is relatively vast. But let's make an attempt to form a brief insight on the systematic establishment of the laws and consequences regarding the black-body radiation.

The Scottish experimental physicist, Balfour Stewart was the first scientist who first hinted at the concept that later came to be known as the black-body Radiation. Balfour Stewart, in1858, conducted an experiment on the basis of which he eventually described as a kind of emission of thermal radiation, which can perform reflection and refraction in terms of rays that obey the Helmholtz Reciprocity Principle that describes the procedure of a ray of light and its reverse ray linked up with facets like reflection, refraction, etc. in an entirely static and non-linear system. Additionally, one needs to know that the mathematical side of this principle is also applied in the mainstream quantum mechanics or specifically in the local transformations referred by the symmetric system within the field of 'T-Symmetry' or Time-reversal symmetry which implies the entropy conservation; though it's not applicable to the macroscopic scenario due to it specified range in the laws of quantum mechanics. In this experiment of Stewart, he didn't mention a lot of significant things like what exactly is the characteristic of that ray such as its wavelength, spectral regime, etc. There was even no correlation of thermodynamically stable, unstable or any kind of system and thus, in some ways, the work of Stewart wasn't complete because there were certain blanks in his theoretical and experimental explanations which became a big problem due to the profound nature of the theory.

On the other hand, Gustav Robert Kirchhoff had no idea about the ideology of Balfour Stewart and as he was already investigating the phenomena in 1859, which was only a year after the Stewart's work came out; he came up with his analogy of spectral wavelength and the emission of visible light. In the whole account of the theory, he didn't mentioned the laws of thermodynamics, but later in one of his books he showed that the theory has a relation with Carnot's rule which stems from and is the second law of thermodynamics. Apart from this, Kirchhoff's other direct contribution in the development of thermal physics through this notion of the black-body is the dependency showed by him from the temperature difference between the absorber and emitter on the dark & bright lines in spectrum. There is another notion defined by Kirchhoff is the "perfect black-body" which lead to his another principle explicated in order to make the "black-body" explanation more intense. This principle known as the Kirchhoff's law of Thermal Radiation simply relates the absorption and emission as equal at the time of heat transfer in the thermodynamic equilibrium. The legacies of Gustav Robert Kirchhoff made a great impact on sciences that even the great physicist Lord Rayleigh's famous "Rayleigh-Jeans law" describes the spectrum of black-body quite perfectly and this law, combined with the interpretation of Kirchhoff, exclusively help in the foundation of Max Planck's theory which led to the development of the first ever fundamental initiation of quantum mechanics. Even one of Boltzmann's contributions, the Stefan—Boltzmann law also plays a very important role in Planck's view on black-body radiation. This law is about the fundamental analogy of radiated power in terms of its temperature. The mathematical side of Stefan—Boltzmann law also plays a vital role in the mathematical formulation of Planck's theory of black-body radiation and hence, the black-body radiation played a significant role in the origin of quantum theory.

Now, finally, it's about the heart and foundational principle of quantum theory, i.e. one of the most significant legacies developed by Max Planck which is named as Planck's Law. This law is fundamentally an explanation of the black-body radiation, which gave rise to a major contribution in the field. Foundationally, it uses Boltzmann's statistical techniques and then mathematically explains the radiant's spectral-energy distribution emitted by a black-body at two ideal conditions; one at the thermal equilibrium and another at no flow of energy or matter between the body and its surroundings. Although there are two more primitive points related to Planck's Law, one is already mentioned in one of the previous, Rayleigh-Jeans law section, which is a formulation from the view-point of classical argument and deals with the intensity of electromagnetic radiation by a black-body as the function of wavelength imposed at certain temperature; and another is 'Wien's Approximation' which is used to describe the spectrum of thermal radiation and more profoundly the short wavelengths. Thus, in the limit of long wavelength or low frequencies, Planck's law becomes the Rayleigh-Jeans law, and in the limit of small wavelength or high frequencies, Planck's law becomes the Wien's Approximation. The whole thing happens due to the change in mathematical formulations; in fact it is said that Max Planck deliberately led this aspect so that the law could be understood in a certain existing domain because both these laws were available at that time.

But this is not that exact point where quantum physics took shape, although it's the foundation this is still not the exact thing that can be credited as the first ever theory regarding quantum mechanics. Actually, ultimately Planck's law regarding black-body radiation contributed to Einstein's concept of 'Quanta' lugging the linear momentum, which became a rudimentary basis for the enlargement of quantum physics. But despite that, mainly Planck's idea is considered to be the basis for quantum theory and it's absolutely true. But, I think it's a tie between Max Planck and Albert Einstein when it comes to "Father of Quantum Physics". But if I were to give an independent opinion then it definitely is Albert Einstein because he was the man who realized the principle and insights of Max Planck, rewriting the laws of physics in a different era.

After a few years of Planck's Law, Indian physicist Satyendra Nath Bose came up with a wonderful idea to describe incomprehensible or non-interacting particles as an occupying space at the thermodynamic equilibrium in the same set of existed discrete energy states. These particles were termed as "Bosons", which don't follow the restrictions provided by Wolfgang Pauli in his famous Pauli Exclusion Principle that fundamentally says that two or more identical 'Fermions' can't occupy the same quantum state simultaneously within the domain of that particular quantum system, and for electron, it postulates that in an atom or molecule (or sometimes called as the poly-electron atom) the same four electronic quantum numbers can't be held by two electrons. Later on, this idea got Einstein's support and now the whole mechanism of this is called Bose-Einstein Statistics. The point is that the statistical mechanics of Bose also provide a theoretical proof of Planck's Law. Albert Einstein also contributed in the account of black-body Radiation with a mathematical framework which is called 'Einstein Coefficients', which helps in measuring the probability of emission of light by an atom or molecule. The Einstein Coefficients relates spontaneous emission and the absorption of light.

Apart from quantum physics, the black-body radiation has been successfully contributed in a lot of remarkable principalities of physics like in astrophysics, cosmology, particle physics, etc. and has proven enormously helpful for not only theoretical sciences but for the practical life as well which is an extraordinary milestone. Even the cosmic microwave background radiation which is an obscure cosmic background radiation that fills all space and has an intense connection with the big-bang is the most perfect black-body radiation ever observed because this is a kind of electromagnetic radiation, which is a residue of early universe, way back to the big-bang and that's why it's also known as Relic Radiation.

18
The Rise of Statistical Mechanics

Bring forward what is true, Write it so that it is clear, Defend it to your last breath. - Ludwig Boltzmann

In the field of quantum physics, there are certain subsets to it which played a very significant role in the development of its ideology in a modernized locution and one of these terms is the statistical mechanics. It is one of the pillars of modern physics. Sometimes, there are certain physical systems or interactions in the domain of quantum physics which obviously has peculiar degrees of freedom; in that case statistical mechanics comes into consideration in order to analyse the respective aspects with certain accuracy which is quite satisfactory as compared to any other techniques. Although, statistical mechanics has specifically certain formulated techniques in its range, it not only applies to quantum physics but also in other parts of theoretical physics and sciences like mathematics, economics, computer science, etc. because there are peculiar methods and formulations which are in statistical mechanics that plays a vital role in several dimensions of sciences in deep-rooted mathematical ways. But because right now we are taking quantum physics in our domain of study, we'll focus on the significance and evolution of statistical methods in quantum physics only which lead the quantum world to become more precisely analysable or predictable.

The initial stage or thought in the direction of statistical mechanics was conceived with the Kinetic theory of gases whose founding father was the great physicist and mathematician Daniel Bernoulli, who is perhaps best known for his contribution in fluid dynamics as the Bernoulli Principle, which is a fundamental establishment of an analogy between the static pressure and potential energy of a fluid. Apart from physics and mathematics, Bernoulli also had a great impact on economics too. There is no question about the fact that Bernoulli was the man who had the primordial insight into statistical mechanics, but the term was coined after a lot of methods came up in the field and the term was used for the first time by an American scientist Josiah Willard Gibbs who had some very thoughtful and fundamental contributions in physics, mathematics and physical chemistry. He coined term in the year 1884, after he gave a clearer direction with his book titled "Elementary Principles in Statistical Mechanics", which came out in 1902, almost eighteen years after the nomenclature of another form of science which we can also call advanced version of Probability Theory. This book is considered as a significant foundation of modern phase of statistical mechanics. From the initial point of view, the most common thing about both quantum mechanics and statistical mechanics is that both of them have consequence as uncertainties; like in the book of Gibbs, he shows the laws of thermodynamics would extensively arise as a subset of peculiar classical system of laws if the intrinsic uncertainty is conceded. Even up to the current time, the work of Gibbs didn't lose its natural elegance including the book which successfully explored a lot of significant aspects about the statistical thermodynamics and statistical mechanics respectively.

But up to the time of Gibbs, there were certain formulations and theories that already existed which indicated towards statistical mechanics and statistical thermodynamics. People like Ludwig Eduard Boltzmann, Rudolf Julius Emanuel Clausius, and James Clerk Maxwell had already defined some very useful insights into such kind of techniques. Rudolf Clausius was the pioneer of thermodynamics who defined both the first and second laws of thermodynamics and the physicist who revived the idea of Daniel Bernoulli about gases consisting of molecules in motion capable of moving into all directions. The term 'entropy' was also coined by Rudolf, which is one of the extensive properties of thermodynamics, and second law of thermodynamics combined with the conceptualization of entropy also played a significant role in figuring out certain very fundamental as well as advanced versions or frameworks to the understanding of the evolvement of universe, space-time curvature etc. It also plays a very fundamental role in conformal cyclic cosmology, which is a framework obtained as a subset of Einstein's Theory of General Relativity, which gives multiple cycles of conformal components as space-like or time-like geodesics and beyond under certain specific realm of mathematical formulations, for example Friedmann–Lemaître–Robertson–Walker metric (FLRW metric), pseudo-Riemannian manifold, etc. Well, perhaps it would be tough for a lot of readers to understand it but I'll try to be as simple as I can to give you a rough idea of what exactly the FLRW metric & pseudo-Riemannian manifolds mean. So, a FLRW metric is about an expanding Universe that is homogenous and isotropic holistically mean the identical expansion of universe everywhere which is the foundational cosmological principle as well, described as a mathematical solution to Einstein's field equations; the pseudo-Riemannian manifold is a realm of differentiable frame towards neighbourhood points with metric tensor (in simple terms, a metric tensor is one of the measuring property of geodesics, and geodesics represent the same aspect of a straight line to the curved space-time), which is arbitrarily bilinear in that peculiar topological space.

So that's the importance of Rudolf's contributions to physics, which also plays a vital role in other parts of theoretical physics as well. But the main point is Rudolf Clausius made fundamental contributions in the study of diffusion of molecules which inspired James Clerk Maxwell to jump into the regime of a kind of new aspect to the probability theory, in the language of physics which later became statistical mechanics. This work of Maxwell is named as the Maxwell speed distribution, which is a continuous probability distribution theory regarding the motion of freely suspended oscillating particles, excluding the interaction between them and is also responsible for the first ever scientific argument that properties of particles, e.g. temperature, can have an equilibrium tendency to be attained and this is the first law of statistical mechanics so far. He also contributed to thermodynamics, but he thought that the second law of thermodynamics can be hypothetically violated or have some vital limitations which he supported with his thought experiment occurring due to the faster movements of molecules which ultimately somehow leads with the disintegration of entropy in the system he called 'Maxwell's Demon'.

Maxwell himself defined the idea in the following words in his book "Theory of Heat":

".... But if we conceive of a being whose faculties are so sharpened that he can follow every molecule in its course, such a being, whose attributes are as essentially finite as our own, would be able to do what is impossible to us. For we have seen that molecules in a vessel full of air at uniform temperature are moving with velocities by no means uniform, though the mean velocity of any great number of them, arbitrarily selected, is almost exactly uniform. Now let us suppose that such a vessel is divided into two portions, *A* and *B*, by a division in which there is a small hole, and that a being, who can see the individual molecules, opens and closes this hole, so as to allow only the swifter molecules to pass from A to B, and only the slower molecules to pass from B to A. He will thus, without expenditure of work, raise the temperature of B and lower that of A, in contradiction to the second law of thermodynamics.

This is the only one of the instances in which conclusions which we have drawn from our experience of bodies consisting of immense number of molecules may be found not to be applicable to the more delicate observations which we may suppose made by one who can perceive and handle the individual molecule which we deal with only in large masses".

"Maxwell's Demon" thought experiment is still getting respect by a lot of eminent scientists. There are also a set of four mathematical equations that are called as "Maxwell Relations" which relates some of the native thermodynamic entities, i.e. temperature, entropy, pressure and volume simultaneously as the range of thermodynamic potentials.

At the same time but at a different place, there was another revolutionary physicist growing up who was highly influenced by Maxwell's work and greatly inspired by him; this person was Ludwig Eduard Boltzmann. He had several revolutionary ideas, which had a great hand in the establishment of statistical mechanics and statistical thermodynamics like Boltzmann Transport Equation (defines the statistical comportment of thermodynamic system in the non-equilibrium state), Boltzmann Distribution (defines the systematic probability distribution of energy by distinguishable and identical particles), etc. The Boltzmann Transport Equation has an automatic consequence as one of his own ideas which is named as 'H-Theorem' which fundamentally explains the decrement tendency of the relative densities i.e. specifically concerns with the distribution function of kinetic energy of molecules in an isolated ideal gas system. This H-Theorem itself has a noteworthy place in the modernization of statistical mechanics.

Boltzmann's contributions later inspired Max Planck too in the form two of formulations named Stefan-Boltzmann law and Boltzmann Constant, which became one of the key reasons behind the birth of Planck's interpretation on black-body radiation.

After these classical people, a lot of work was established in the field of statistical mechanics by some great minds like Albert Einstein, Paul Dirac, etc. This whole account of statistical mechanics and statistical thermodynamics having very core applications in the various aspects of sciences like Mathematics (probability theory, differential equations, etc.), theoretical physics (e.g. quantum physics, particle physics etc.) and economics (e.g. financial mathematics, engineering economics etc.), etc., and with these extraordinary applications, sciences were raised to a whole different level of imagination with highly profound characteristics.

Part 5:
Einstein's Annus Mirabilis & Beyond

19
The Photoelectric Effect

Two things are infinite: the universe and human stupidity; and I'm not sure about the universe.
- Albert Einstein

The phenomenon of Photoelectric effect was first introduced by two German physicists, i.e. Heinrich Rudolf Hertz and Wilhelm Ludwig Franz Hallwachs, in 1887. Though H R Hertz is more often regarded as the father of the photoelectric Effect, Hertz first observed the charged object intensively losing its charge when irradiated with ultraviolet rays (or UV rays). Also, he noticed the negative discharging of ultraviolet light with the creation of rays similar to cathode rays. Then in 1900, this concept was generalized further and extended by another German physicist Philipp Eduard Anton von Lenard with the discovery of ionization of gases from ultraviolet rays and thus held the experimental realization accordingly. The concept of photoelectric emission also came into consideration with the use of Philipp Lenard's work when he stated that the energy ejection of any electron from a cathode doesn't depend on the intensity of the incident light but on the wavelength only as he observed that energy of an individual emitted electron increases with the frequency of light in the spectrum which concerns corresponding colours.

Apart from Hertz, Hallwachs and Philipp Lenard, British physicist Joseph John Thomson's investigation on ultraviolet light in the Geissler tube (a gas-filled tube generally used to demonstrate the principles regarding electric glow discharge) is also responsible for the development of the photoelectric effect in early days. Thomson defined that the ejected particles were identical, as commonly found in the case of cathode ray. He called these particles 'corpuscles', which were later named electrons. Before all of them, the French physicist Alexandre Edmond Becquerel came up with a chemical and physical phenomenon which describes the production of electric current and voltage in a material attributed on the vulnerability of light. He determined this aspect of physics in 1839 and successfully established the analogy between electronic characteristics of any material and light, which later became helpful in the determination of the photoelectric effect. The analogy discovered by Edmond Becquerel is known as Photovoltaic Effect, which is closely related to the photoelectric effect, and specifically used in the excited state of electron as any other charge-carrier when light is absorbed. Edmond Becquerel used an electrochemical cell to demonstrate this. On a side note, one should know that an electrochemical cell is a device that is used to produce electrical energy from chemical energy or uses the electrical energy to perform chemical reactions.

Now, coming to the further development in this regime; Einstein came up after the theory of Philipp Lenard about the photoelectric effect, and proposed the quantization of light as the formation of discrete packets called 'quanta' which later came to be called as the 'photon around 1926 by the American physical chemist Gilbert Newton Lewis. Einstein not only made the emission of free charge carriers more precise with the conceptualization of photoelectric effect in terms of mathematical formulation, but also made Planck's law more profound regarding the black-body radiation. The mathematical equation of photoelectric effect, which he proposed in his paper titled "On a Heuristic Viewpoint Concerning the Production and Transformation of Light" showed that energy in each quanta of light is equal to the frequency multiplied by a constant which later came to be called as Planck 's constant. Einstein's prediction about the absorption of quanta was confirmed after a few years of his discovery, around 1915, when American experimental physicist Robert Andrews Millikan verified it with his famous 'Oil-drop experiment' that got him a Nobel Prize "for his work on the elementary charge of electricity and on the photoelectric effect". The work of Albert Einstein became the foundation for quantum mechanics, which now has vital application is many fields of physics like electrostatic levitation (by the creation of electric field in application by Millikan's experiment), spacecrafts (by the production of static charge from the effect), photoelectron spectroscopy (due to the consideration of binding energy and work-function in the effect), etc. Thus, Einstein's most initial discoveries too created a major impact on sciences and humanity.

20
The Brownian Motion Theory

These motions were such as to satisfy me, after frequently repeated observation, that they arose neither from currents in the fluid, nor from its gradual evaporation, but belonged to the particle itself. – Robert Brown

Brownian Motion is basically the random movement of particles suspended into the fluid due to collisions with other surrounding particles with respect to the surrounded medium of molecules. The Brownian Motion Theory was first defined or theorized by the Scottish botanist Robert Brown around 1827, nearly centred around the beginning of quantum ideology. Although there are some arguments about the discovery of Brownian motion despite of it being named after Robert Brown, it's sometimes said that the motion was first theorized by Dutch scientist Antony van Leeuwenhoek who was a self-taught biologist, also regarded as "The father of microbiology". Well, the data we have is certainly conflicting, but most of the times the observations of Brownian motion are attributed to Robert Brown only; though the work of Antony van Leeuwenhoek influenced Brown's curiosity a lot.

The writings of Robert Brown himself point out the following aspects of his observation and beyond :

> "These smaller particles, or Molecules as I shall term them, when first seen, I considered to be some of the cylindrical particles swimming vertically in the fluid. But frequent and careful examination lessened my confidence in this supposition; and on continuing to observe them until the water had entirely evaporated, both the cylindrical particles and spherical molecules were found on the stage of the microscope.
>
> In extending my observations to many other plants of the same natural family, namely "Onagrariæ", the same general form and similar motions of particles were ascertained to exist, especially in the various species of "Œnothera", which I examined. I found also in their grains of pollen taken from the "antheræ" immediately after bursting, a manifest reduction in the proportion of the cylindrical or oblong particles, and a corresponding increase in that of the molecules, in a less remarkable degree, however, than in "Clarckia".
>
> Grains of pollen of the same plant taken from "antheræ" immediately after bursting, contained similar sub-cylindrical particles, in reduced numbers however, and mixed with other particles, at least as

> numerous, of much smaller size, apparently spherical, and in rapid oscillatory motion.
>
> Extremely minute particles of solid matter, weather obtained from organic or inorganic substances, when suspended in pure water, or in some other aqueous fluids, exhibit motions for which I am unable to account."

It can be clearly seen that in the last point mentioned in the writings of Robert Brown, he is facing certain difficulty in explaining the motion clearly; but this made his contemporaries and upcoming scientists think over it. It's not exactly about the legacy of Robert Brown or Antony van Leeuwenhoek because if we go deeply into ancient philosophy then the Roman philosopher and poet Titus Lucretius Carus also somehow defines the movements of such particles as referred in the theory of Brownian motion. The famous "Book 2", which is a version of Lucretius' book "De rerum natura" or "On the Nature of Things", that specifically concerns with the motion and other related aspects of atoms along with their combinations in natural abstract, describes the phenomenon of such motion in the below lines as (extracted from the English version of his original content) :

> ".....Nor was the store of matter ever more closely packed nor again set at larger distances apart. For neither does anything come to increase it nor pass away from it. Wherefore the bodies of the first-beginnings in the ages past moved with the same motion as now, and hereafter will be borne on for ever in the same way; such things as have been wont to come to being will be brought to birth under the same law, will exist and grow and be strong and

lusty, inasmuch as is granted to each by the ordinances of nature. Nor can any force change the sum of things; for neither is there anything outside, into which any kind of matter may escape from the universe, nor whence new forces can arise and burst into the universe and change the whole nature of things and alter its motions.

.....Observe what happens when sunbeams are admitted into a building and shed light on its shadowy places. You will see a multitude of tiny particles mingling in a multitude of ways... their dancing is an actual indication of underlying movements of matter that are hidden from our sight.

.....It originates with the atoms which move of themselves. Then those small compound bodies that are least removed from the impetus of the atoms are set in motion by the impact of their invisible blows and in turn cannon against slightly larger bodies. So the movement mounts up from the atoms and gradually emerges to the level of our senses so that those bodies are in motion that we see in sunbeams, moved by blows that remain invisible.

.....Nor need we wonder; for since there is so great a store of them, that neither have they any limit, as I have shown, nor any sum, it must needs be, we may be sure, that they are not all of equal bulk nor possessed of the same shape.

......Those things indeed must be made of particles more round and smooth, which are liquid with a fluid body: for indeed a handful of poppy-seed moves easily just as a draught of water; for the several round particles are not checked one by the other, and when struck, it will roll downhill just like water.

Hence, it proves that the ideology of Brownian (like) motion theory is quite elegant right since ancient times. The modernization of this theory is mainly attributed to Albert Einstein and with the help his formulations, existence of molecules or atoms came into account with the corresponding mathematics; Einstein's paper titled "On the movement of small particles suspended in a stationary liquid demanded by the molecular kinetic theory of heat" on the Brownian motion theory is strictly depends on an intuitive mathematical framework that became an initial point of an entirely profound field to study. Although, Danish Astronomer Thorvald Nicolai Thiele was the first person who made a fundamental attempt to decode the mathematics behind Brownian motion, but it didn't turn out to be extremely profound because it was just based on the existing mathematics at that point of time like the regression analysis which has a standard method of Least Squares to approach the solution of over-dominant system. Despite of all these aspects, great philosophers like Leucippus and Democritus believed on the discrete behaviour of the matter constituents; the English physicist and chemist John Dalton postulated the existence of atom around 1808 through his atomic theory and eventually by the end of nineteenth century great Physicists and Mathematicians like Boltzmann, Clausis and Maxwell offered some path-breaking works in statistical mechanics, specifically the molecular kinetic theory of gases. But then also existence of atom and molecules didn't receive general acceptance. For instance, scientists like Ernst Waldfried

Josef Wenzel Mach (or Ernst Mach) and Friedrich Wilhelm Ostwald remained sceptical about the existence of atoms.

Before Einstein, there were a few people who had tried to explain the observation or description of Robert Brown, for example Italian physicist Giovanni Cantoni came up with the random motion of Brownian particles as a cause of collisions with the molecules of the described fluid; but the entire scenario described by Giovanni was also not proven very successful due to some arguments by contemporary people regarding linear collisions. In-fact he himself was familiar with a few of explanation towards the motion by his contemporaries scientists, but none of them were proven satisfactory.

But, then in year 1905 Einstein's theory on Brownian motion explained the phenomena correctly, but it was not the only paper which was constructed on this phenomenon because according to some science historians, his core work in the doctoral thesis which was named as "A New Determination of Molecular Dimensions" extended the Theory of Brownian Motion in certain degree of excellence as compared to that of before presented ever in terms of hydrodynamics, which also plays a vital role as an essential framework to his formulations on the core Brownian Motion because it gives several insights over it, like methods to measure the atomic dimensions, describing the Avogadro's number by the consideration of large molecules (specifically; sugar molecules) in solution which solved the most fundamental resistance towards his Theory of Solution concerning semi-permeable membrane (it is a polymeric membrane which uses diffusion for allowing the flow of certain peculiar ions or molecules through it) or osmotic pressure (it is that essential amount of pressure which has to be applied to a solution in order to intercept the inward glide of corresponding pure solvent over semi-permeable membrane). But ultimately, almost all account of Brownian Motion Theory with respect to Einstein's statistical methods have been taken into a singular consideration with his book which was published around 1926 titled "Investigations on the theory of Brownian Movements". There is a huge misconception regarding Einstein's framework of Brownian Motion Theory, that he developed the doctrine only for the profound evidence of atomism and kinetic theory of gases; but the fact is, he had developed the interpretations of statistical mechanics independently during 1902-1904 and based on them developed his original interpretation of the Brownian motion theory, which concerns a bigger regime, i.e. beyond just kinetic theory of gases, and including arbitrary interactions of physical systems at the microscopic level. Einstein also wrote about it in a letter to his friend Marcel Grossmann in 1904 who was working on the non-Euclidean geometry:

"There is a remarkable similarity between us... You treat geometry without the parallel axiom; and I treat the atomistic theory of heat without the kinetic hypothesis".

Thus, Einstein's interpretation over the relative mathematics was beyond that of Boltzmann. But, yes, Boltzmann's foundational aspects has been very helpful to many physicists and are relevant up to the current time with his theories or formulations finding more and more applications in the enhancement of sciences. But, as far as Einstein is concerned, his perception on Boltzmann's statistical methods were quite different and Einstein disagreed with Boltzmann's interpretation because according to Einstein, the system has to be defined in the deterministic domain of the respective dynamics even in the system of equilibrium. With the help of Einstein's investigation on the theory, many great formulations have been made in the area of statistical thermodynamics too that have helped a few physicists develop the theory of Brownian motion in a more advance form, for example the Theory of Rotational Brownian Motion by physicist and physical chemist Peter Joseph William Debye who used Einstein's formulations to obtain the distinction between the chaotic patterns or orientations of polar molecules giving rise to the permanent electric dipole, which became one of the vital aspects for the studies concerning dielectric materials. Apart from Albert Einstein, physicist Marian Smoluchowski also came up with his independent explanation on Brownian motion almost a year after Einstein. Both of them brought up results independently, which were mostly similar to one and the other, but Einstein's formulations were still more profound because they independently dealt with a lot of different statistical realms including thermodynamics, hydrodynamics, etc. in a generalized way. Though Marian Smoluchowski's model also starts from the same point as that of Einstein's, which is probability distribution with respect to the displacement of a Brownian particle along a peculiar axis or direction in terms of mathematical attribution of time, but the system to get on the result was different from one and the other. Apart from this, the identical thing among both of them was the ultimate conclusion that the Brownian motion of particles has certainly measurable sizes concerning the expression of movements of atom in a fluid. Then, in 1908, French physicist Paul Langevin came forth with his research in the field, which he regarded as "infinitely more simple than Einstein's" because Langevin fundamentally used Newton's second law of motion to further formulate his insights whereas on the other hand Einstein began with a hypothesis and then preceded original mathematics on the basis of that hypothesis. The most foundational work of Paul Langevin on the development of Brownian motion theory is the Langevin Dynamics, which is concerned with the figuring out of degrees of freedom using stochastic differential equations, which are the mathematical prerequisites to mainly describe the stochastic process (it's a random process concerned with the corresponding suitable mathematical distributions). The Langevin's framework is also constructed on the underlying formulations of Markovian Process, which also established under the stochastic process as defined previously.

But, ultimately, both the works of Einstein and Langevin come to a common ground, which can lead the formation of both results regarding Einstein and Langevin respectively and that formulation is Fokker–Planck equation that defines the change of state with time or time evolution of the probability density as the function of velocity underlying the identical ground of random forces described in Brownian motion theory. When this Fokker–Planck equation leads with the position function of particle then it becomes "Smoluchowski equation" which extends the theory in a significant way. Hence all the explanations regarding the theory of Brownian motion ultimately can be interlinked with certain mathematical formulations.

Albert Einstein's interpretation came into physical existence with the experimental confirmation conducted by Jean Baptiste Perrin around 1908, who got the Nobel Prize in Physics "for his work on the discontinuous structure of matter, and especially for his discovery of sedimentation equilibrium." His observation was stimulated with Einstein's statement about the thermal molecular motion that could be easily observed under a microscope according to him and ultimately Perrin's conclusion from the experiment which verified the existence of atoms impressed Einstein too because the observations were quite precise. Hence, Einstein wrote a letter to Perrin saying, "I did not believe that it was possible to study the Brownian motion with such a precision".

The theory of Brownian motion with its formulations has been very useful to various fields of physics, mathematics and economics, etc. in developing many revolutionary insights. For example, French mathematician Louis Jean Baptiste Alphonse Bachelier became the first person who imposed stochastic process in Mathematical Finance in his doctoral thesis "The Theory of Speculation" in 1900, which is often considered as the birth year of mathematical finance.

21
Bose-Einstein Analogies

The most beautiful thing we can experience is the mysterious. It is the source of all true art and science.
- Albert Einstein

Statistical mechanics has been an active area of research right from the time when people like Maxwell, Boltzmann and Einstein investigated on it with different terminologies. Some of the modern interpretations in this area were put forward by several great frameworks like the Fermi-Dirac statistics which has already mentioned previously; but there is another contemporary execution of statistical mechanics which has a revolutionary role in the quantum world, in its own unique way and extends the insights of analytical subatomic realm in the form of many new introductions and methodologies. Around 1924, a young enthusiastic scientist from India wrote a letter to a famous scientist, requesting him to review his idea and offer his opinion on it; the letter was also sought a translation into German language and proceed for publication if the idea was convincing enough. This Indian scientist was Satyendra Nath Bose writing to Albert Einstein. It's seemed very strange that someone belonging to a completely different picture, writes to one of a most famous person seeking translation of his manuscript; but it is true, and Einstein eventually wrote back informing him that he has translated the papers and proceeded as per the desire of S. N. Bose. Interestingly, it is said that Bose translated Einstein's Theory of General Relativity into English even before writing to Einstein. In English, the idea is entitled in "Planck's Law and Hypothesis of Light Quanta", which is now called the Bose—Einstein Statistics. The idea utilizes the mathematics of black-body radiation by Max Planck or the radiations' energy distribution to be precise, and develops a kind of mathematical structure which analyse the photon independent of the classical electrodynamics, and it also devised the thermodynamic probability of a state. However, Einstein found certain discrepancies in Bose's idea regarding Bohr's correspondence principle and a few others, but the overall paper offered some advanced insights in the statistical terms. Although some of the relevant aspects about Bose—Einstein statistics are already introduced in one of the previous chapters regarding the Planck's black-body revelation, in which the notion of bosons are also mentioned, even though the term "Boson", particles that obeys the Bose—Einstein statistics was coined by Paul Dirac. Both the Fermi—Dirac statistics and Bose—Einstein statistics founded the study of Particle statistics. But, the catch in the story is all about Einstein's brilliant take on the idea of Bose and this can be observed with the fact that Einstein extended this idea even more efficiently than Bose himself.

Einstein's comment on the paper of Bose was expressed in following lines:

"Bose's derivation of Planck's formula appears to me to be an important step forward. The method used here gives also the quantum theory of an ideal gas, as I shall show elsewhere".

So, Einstein got the idea of ideal gas from the interpretation of S N Bose and thus he investigated it and got some very extraordinary results although; this conceptualization that obeys Bose-Einstein statistics was then conceived by Bose only, but it was extended up to the regime of massive particles or atoms by Albert Einstein. The idea of Bose is known as 'Bose Gas', but Einstein then observed that the notion of Bose gas in terms of bosons can be generalized with any suitable dependency of any critical low temperature which turned out as the famous term, i.e. "Bose—Einstein Condensate" or "BEC", which is denoted as the fifth state of matter followed by Plasma as fourth state; so, in short, when the gas by bosons cools down it forms the BEC. In Bose gas, the actions of several bosons, especially in the non-interacting phases are termed as 'Photon gas', which significantly developed the comprehension of Planck's law as the black-body radiation more conveniently. The precision towards the system of mass-less bosons were expanded or explained through the well-known notion of 'Debye model', which is attributed to brilliant physicist Peter Joseph William Debye and this gave rise to the concept of 'phonon', which also contributes to the field of condensed matter physics very effectively as it focuses on the quasi-particles as basic characteristic. The formulations in the ideal Bose gas utilizes some underlying mathematical descriptions of statistical mechanics, for example, Grand Canonical assemble, which defines the properties of particles and respective states at the thermodynamically equilibrium extremity, such that the idea is natively connected to various different aspects of thermodynamics like the 'Sackur–Tetrode equation' that is concerned with the classical form of ideal monatomic gas and entropy, which is named after two eminent scientists and contemporaries of Einstein - Otto Sackur and Hugo Martin Tetrode respectively.

Apart from that, in the one-dimensional quantum framework, Bose gas is also related to another formulation which is noted as the one-dimensional non-linear version to the Schrödinger's equation; in-fact both possesses equivalence in the one-dimensional regime. There are some other fundamental set of rules and mathematical fabrications which are very basic to both Bose—Einstein Statistics and BEC or Bose gas, like when the matter comes to several identical bosons and quantum coherence concerning the phenomena, then there are some sort of tools that work to deal with it and these are known as "Bose—Einstein Correlations", which uses some other aspects of quantum mechanics, and even quantum optics that has to do with radiant energy or more precisely the formulation called "Hanbury Brown and Twiss effect" or HBT effect; similarly there are some methodologies in Fermi-Dirac Statistics too. All the mathematical extensions regarding the ideal Bose gas or Bose—Einstein Condensate has a fundamental representation that relates a certain critical temperature and phase transition with respect to the chemical potential. The entire conceptualization of both Bose–Einstein statistics and Fermi–Dirac statistics are highly remarkable, however there are some sort of alternatives too in the field of statistical mechanics whose compilation is called the Parastatistics as they hold some profound quantum frameworks that achieve similar degrees of excellence in order to predict or calculate the statistical aspects with the respective techniques for example, the generalization of both fermions and bosons is known as the Braid Groups, and the whole systematic notion is called as Braid Statistics. So, in this way Albert Einstein endorsed the scientific discovery of a young brilliant scientist and gracefully gave humanity some very fundamental aspects of sciences that laid the foundation stone for many branches of quantum physics combined with statistical mechanics.

22
Bohr—Einstein Debates

Prediction is very difficult, especially if it's about the future. - Niels Bohr

In 1927, the fifth Solvay Conference was announced in Brussels, which turned out as the most astonishing conference ever because, all the best minds of science participated in it like Albert Einstein, Max Planck, Hendrik Lorentz, Paul Langevin, Max Born, Niels Bohr, Paul Dirac, Marie Curie, Wolfgang Pauli, Werner Heisenberg, etc. The conference meant to discuss quantum mechanics. But the conference got enormously famous only due to the presence of Einstein and Bohr. It is said that initially Einstein didn't want to speak on the subject because it was famously known that Einstein's opinion on quantum mechanics was quite different compared to the rest of the scientists working in the field. But, when Heisenberg and Born's turn came and they expressed their firm opinion that the explanations of quantum mechanics had reached a level where no modifications were required to make, Einstein did not like this behaviour of the two prominent scientists but he remained silent, and then Niels Bohr spoke on the subject and proposed almost same thing and this eventually hurt Einstein's thought that was sceptical of the profound nature of quantum mechanics, which had been finally coming up with its probabilistic principles. Another thing that Einstein didn't like was their attitude towards the subject, which was clearly suggesting that quantum mechanics was the most accurate theory of nature and needed no further discoveries because the field was at its end. Thus Einstein stood up and played his master stroke in the conference in the form of his thought experiment. In the conference Einstein attacked Heisenberg's 'Uncertainty Principle' or 'Indeterminacy Principle' and the Bohr's 'Complementarity Principle; these are two of the most fundamental principles which played a key role in the understanding of quantum mechanics as the study of possibilities in nature. The Bohr—Einstein Debates, if understood in a simple way, goes somewhat like this - Einstein gives some astonishing thought experiments whose works are to contradict these two principles and similarly, Bohr had to destroy them with rational reasoning to keep his understanding of quantum mechanics alive. So, at the very first thought experiment; Einstein used the concept of conservation of momentum and conservation of energy via the double-slit experiment in order to violate the uncertainty principle, his explanation of imagination goes like this:

"So make a foundational assumption that in between the wall of double-slit and the medium of light, there is another wall of single slit placed in such a way that when the beam of light fired via that single slit, the beam must be treated as the form of wave similarly like in the primordial experiment and when the two wave forms passes through the second wall of double-slit then both will mutually interfere and thus produce the interference which we observe. But, as the light also has particle-like behaviour too and thus the only deflection possibility of that particle-like aspect is the ricocheting with respect to the very first wall of single-slit. Now, suppose that the first wall is on a movable roller then we can predict the slit from second wall from which the photon will pass due to the deflection of first wall by the bouncing-off photon with the laws of conservation of momentum; but it's just impossible because the wave-particle duality can be simultaneously occur in both forms and due to the wave-function collapse, ultimately we won't be able to predict the previous case which is regarding the tracing of the pathway of photon which is going to pass through any particular slit from the two. But, the condition only satisfies if the wave-like and particle-like behaviour takes place simultaneously with respect to their measurements and this violates the Heisenberg's uncertainty principle".

But, on the other hand, Niels Bohr was there with a solution which says that if it has been assumed that the photon is striking the first slit leading its movement, thus we can't know the position of slit and its movement before the photon incident, because the photon itself is a quantum entity so the indeterminacy comes into consideration; hence the thought experiment of Einstein just changed the location of Uncertainty from one to another.

The entire argument of Bohr proved that Einstein's explanation is only applicable if the whole concept of Uncertainty is overlooked, which is not possible in the quantum realm due to its experimental evidence. But still Einstein wasn't fully convinced about Bohr's solution because Einstein was an admirer of classical methodologies, which he supported instead of modern aspects especially with respect to quantum mechanics.

The conference was finally over, but when three years later the same conference was announced the same people were again there, this time Einstein was better prepared to diminish the phenomena of possibilities in the quantum realm. Einstein proposed another thought experiment which went this way –

"Suppose a box with perfect reflective mirror walls and inside that box, there is light bouncing around combined with a clock and now note the weight of the box. After certain time, the clock will be responsible for opening a gateway for one photon to skip the box at the peripheral side of the box or conveniently symmetric to the clock at a peculiar point of time and now again note the weight of box; due to the distinction of mass, you can measure the mass of the photon and if you know the mass of anything, you can calculate its energy by Einstein's famous Mass-Energy Equivalence. Hence, with this, one can measure both Energy and Time simultaneously which violates both Uncertainty Principle and Complementarity Principle because apart from Position and Momentum, Energy and time also reflected in the same way in these two principles".

The "Box" involved in this thought experiment is called Einstein's Box or sometimes as 'Einstein's Light Box'. This hypothesis of Einstein had Bohr anxious and awestruck; it is said that Belgian Physicist Léon Rosenfeld who was also a part of the conference expressed Bohr's reaction in the following words,

"It was a real shock for Bohr; who, at first, could not think of a solution. For the entire evening he was extremely agitated, and he continued passing from one scientist to another, seeking to persuade them that it could not be the case, that it would have been the end of physics if Einstein were right; but he couldn't come up with any way to resolve the paradox. I will never forget the image of the two antagonists as they left the club...."

But eventually, Bohr got a solution for this. Next morning he resolved the "Einstein's box" with one of Einstein's own legacy. Bohr explained that due to the conservation of momentum, the box itself must exhibit aberration after the escaping of photon; now, the question rise here is how much recoil should be there and this thing can't be known accurately because of the Uncertainty Principle. Apart from this, as it generally known that weight has no meaning without Gravity and the box itself is inside the gravitational field. Now, the most accurate description of gravity is attributed to Einstein's Theory of General Relativity which says that gravity is a consequence of the space-time distortion or vice-versa through which, the passage of time is also affected and this is one of the key reasons for the working of Global Positioning System (GPS). Hence, one can't ultimately be sure to know how much the time shown by clock is affected by its movement.

This was the official end of the debate but Einstein still was ultimately in favour of a more generalized theory for quantum mechanics and with the side of modifications required by quantum mechanics. Some of Einstein's other thought experiments and theories also pointed towards the incompleteness of the whole mechanism of theories in quantum mechanics, but later on Bohr ignored all of them because he always wanted to keep his understanding at a safer region of quantum mechanics. Although, the experimental evidences that were brought out in the last few decades clearly favour the current explanations in quantum physics which are quite profound too, but perhaps the possibility exists that a few experiments conducted are presuming the theories of quantum mechanics to be crystal clear like Niels Bohr did, or that some experiments have measurement defects because the quantum theories deals at a very microscopic level and thus gives the results that obey the corresponding explanations like observer effect and other uncertainties. But then also, without any scepticism, the quantum mechanical properties are quite elegant in both theoretical and experimental aspects as previously mentioned at certain places.

23
The EPR – Paradox

Imagination is everything. It is the preview of life's coming attractions. - Albert Einstein

After the Sixth Solvay Conference as previously mentioned in the Bohr—Einstein Debates, Einstein still had this thing in his mind regarding the incomplete nature of quantum mechanics. Actually, there were two phases of quantum mechanics, one corresponding to the theories that came out between the years 1900-1925, which is called Old Quantum Physics and the other phase corresponding to the theories brought out after that period known as Modern Quantum

Physics or New Quantum Physics. The modern views or interpretations in the field included those scientists who are counted as the founders of quantum mechanics after Einstein and Planck. Although it's not that the so-called "Old" and "Modern" had nothing common, but the key thing is that both the eras differ in the way they interpret the interpretations, meaning before or up to 1925, not all; but most of the elucidations were tending to the certainty of that particular phenomenon whereas almost all of these principles constructed before that year, have been turned into a vital consequence of brinkmanship due to the principles or theories defined after the period of 1900-1925.

Einstein specifically had a problem with these modern perceptions of the quantum theory which had nothing to do with the certainty or the deterministic nature of the Universe. Hence, to prove the fragmentary nature of quantum mechanics, he came up with this idea of "EPR Paradox" or Einstein–Podolsky–Rosen Paradox, which is attributed to Albert Einstein, Boris Yakovlevich Podolsky, and Nathan Rosen. The Paradox came out in 1935 with the paper titled "Can Quantum–Mechanical Description of Physical Reality be Considered Complete?". This was in the form of a thought experiment, which proved that the existing account of quantum mechanics was not full or complete.

According to the conclusion of this paradox or thought experiment, Einstein, combined with his two colleagues, clearly showed that the mathematics of wave-function or extendedly, quantum mechanics is incompatible with the physical reality, i.e. what they ultimately described as: "We are thus forced to conclude that the quantum-mechanical description of physical reality given by wave-functions is not complete".

Now the question arises how they got up to this conclusion. So the first thing is mathematics, as they utilized some of the fundamental mathematical formulations like eigenfunction, orthogonal functions etc. (for your information, eigenfunction is nothing but a kind of eigenvector, which is a fundamental linear transformation over any non-zero function that uses a scalar factor called as eigenvalue; or this can be understood as an operator which is generalized as linear form that meant to be in the domain of specific function space subjective to peculiar boundary condition, i.e. give rise to any non-zero function. Similarly, the notion of orthogonal functions is a vector space with a bilinear form constructed under the domain of any particular suitable function space). Apart from mathematics, the idea originated as a profound aftermath of imagination. Before knowing this whole idea, one needs to be familiar with the concept of 'Spin' that started with Wolfgang Pauli when he realized that electrons must obey a rule in order to describe their energy levels in atoms, the rule that turned out as one of the most famous discoveries today known as the "Pauli Exclusion Principle". Particles exhibit certain spins as a consequence of angular momentum in quantum scale and these are noted as 'spin-up' and 'spin-down' respectively; such that if the particle spins clockwise on its axis, then the spin will be counted as 'spin-up' and if the particle spins counter-clockwise, then the spin will be considered as the 'spin-down'. These two components of wave-functions are called the 'spinors'. So now, the ultimate ideology is that if two particles now exists together, but subsequently they are distant or apart, for instance suppose one particle is located in India and another particle has been somehow shifted to Denmark, then despite of the fact that both have now no relation to each other due to so much distance between them, they are linked with each other because mathematically if the particle located in India has 'up' spin then the particle located in Denmark will have 'down' spin or vice-versa. The particles can be apart with a distance of several light years, but the same thing would happen and this phenomenon is now called quantum entanglement, which Einstein called the 'spooky action at a distance and Bohr believed to be an 'entangled pair of particles' with single combined wave-function that possesses the superposition; and thus, as per the Copenhagen Interpretation any measurement with respect to any particle will lead to the collapse of entangled wave-function and this entire phenomena would allow the communication of both particles even back in time, i.e. no limitation of the speed of light. Whereas, Einstein believed in something like the theory of hidden variables, which briefly says that if such a thing is taking place then they're not communicating to each other but rather, they are predetermined with certain information that their state is going to be like this or that way; this information is the constituent of some hidden variables in the very native behaviour of each and every particle. Another big problem in this phenomenon is that the communication requires more than the speed of light for the production of required angular momentum of their spin because spin is nothing but an underlying form of angular momentum carried by particles and atomic nuclei; thus this is certainly impossible due to the Einstein's Theory of Special Relativity and violates the local realism. In the term 'local realism', 'realism' means the universe exists independent of the mind of the observer and the term 'local' comes from the notion of locality which is the idea proposing that the each distinct segment of the universe exclusively acts on its instantaneous surroundings, which is also quite rudimentary to the Einstein's Relativity which indicates that the series of cause and effect (or causality) can't transmit faster than the speed of light. But, if the concept of locality more elaborated then it can't affect the underlying reality of anything which is distant apart i.e. if something occurred at one point of the universe; can't affect anything at a different location of universe which is not quite close or neighbourhood to that incident on an immediate effect or simultaneously.

After all of these things, American physicist and mathematician David Joseph Bohm came up with a thought experiment called the 'EPR—Bohm Thought Experiment' or 'Bohm's Variant', which determines the other aspects that cause the phenomena apart from position and momentum measurements respectively. Bohm also defines his own hidden variable theory named Bohmian mechanics or Pilot Wave Theory or "de Broglie—Bohm Theory", which also express the non-locality of quantum mechanics, or in other words postulates that the reality remains as it is even when unobserved and that quantum mechanics is a non-deterministic theory which doesn't predict the consequences of nature with certainty similar to Einstein's EPR Paradox. Niels Bohr's take on the EPR Paradox was certainly not much exciting though he published a whole paper on this issue in the same year as the EPR was published, i.e. in 1935. As per Bohr, the EPR Paradox is not a good assumption or description to justify the views of all three scientists (Einstein, Podolsky and Rosen) regarding the incompleteness of quantum mechanics. Despite of the fact that Bohr and Einstein never agreed with each other throughout their lives, they had huge respect for each other which maintained the dignity of science in not only their own time but even up to this day and age.

In 1964, there came another remark on the EPR Paradox from an Irish physicist John Stewart Bell, who proposed a mathematical ground to test the paradox. His formulation is famously known as Bell's Theorem or Bell's Inequality, which proves that the account of quantum mechanics is irreconcilable with the local hidden variable theories or simply, that quantum mechanics is a non-deterministic description of universe.

The turn of this smooth story came into early 1980s, when a French physicist Alain Aspect came out with an experiment regarding Einstein's Paradox. He used photon pairs with entangled polarizations instead of focusing on the entangled spin of electron-positron pair. Eventually he got his result in the form of a correlation between the alternatives as the quantification of polarization axis with respect to one photon and terminal polarization direction of its entangled associative, and thus Bell's Inequality violated, and as they are compatible with the EPR Paradox hence guessed that if an entanglement experiment contradicts the Bell Inequality, then local realism is also ruled out. But, here's the catch that violation of Bell's Inequality will only rule out locality without affecting Einstein's Special Relativity and Causality because non-locality and Einstein's Relativity can both be consistent simultaneously due to their more fundamental focus on the Causality as compared to the locality in order to sustain the argument that nothing or no information transmits faster than the speed of light. In fact none of the experiments regarding the quantum entanglement allow any information to be exchanged between any of the particles, one can only observe the effect of entangled pairs after the distinction comparison of corresponding measurements, for example the demonstrations of Quantum Eraser Experiment or specifically the Delayed-Choice Quantum Eraser with respect to the quantum entanglement via the techniques of interferometry.

Hence, Einstein's idea came up with a realistic phenomenon which now holds a lot of applications in several fields; even quantum entanglement and the relative quantum states play a vital role in the enhancement of technology regarding quantum computing. The Copenhagen interpretation is consistent with all experiments conducted in the last few decades; but still the EPR Paradox is one of the centres of attraction for scientists due to a lot of unanswered questions that still remain. But from Einstein's legacy on quantum physics, it can be observed that quantum physics needs certain positive enhancements to make it full of certainties. As it is said, Einstein once asked his colleague Abraham Pais, "Do you really believe the moon is not there when you are not looking at it?". Pais later wrote a biographical book on Albert Einstein and named it "Subtle is the Lord: The Science and the Life of Albert Einstein".

Part 6: The Birth of Modern Quantum Interpretations

24

Niels Bohr – The atomic theory, Bohr-Kramers-Slater theory, Copenhagen interpretation and correspondence principle

If quantum mechanics hasn't profoundly shocked you, you haven't understood it yet. -Niels Bohr

After Albert Einstein, Max Planck and early scientists like Paul Langevin, Marian Smoluchowski, and other eminent scientists came entered the discussion and it came to be the turning point of quantum mechanics because, perhaps, even Einstein himself didn't realized that his deterministic milestone was going to turn into a study and compiling methodologies of endless possibilities. These scientists mainly included Niels Henrik David Bohr (or Niels Bohr), Werner Heisenberg, Arthur Holly Compton, Erwin Schrödinger, etc. All of them are considered as pioneers of quantum mechanics after Einstein and Planck became founding fathers and greatest pioneers of quantum physics. Evidently, their contributions to the subject lead to its development into a higher order of excellence, which has now turned into a whole set of applications and opportunities. So let's briefly discuss some of the key contributions of Niels Bohr who played a vital role in the fundamental enhancement of quantum physics after the above discussed legendary scientists of the quantum-universe.

Niels Bohr (1885-1962) was a Danish physicist who is generally known for his contribution in the atomic theory for which he also got Nobel Prize of Physics in the year 1922, specifically "for his services in the investigation of the structure of atoms and of the radiation emanating from them". In the field of atomic theory he made some of very important contributions, for example, he proposed Bohr model or Rutherford-Bohr model in 1913, which was a systematic construction of tiny and dense nucleus surrounded by orbiting electrons similar to the solar system. The title of Bohr's paper was "On the Constitution of Atoms and Molecules". In this paper, Bohr showed certain mathematical formulations too which were meant to make the theory or model more effective and lead various dimensions of analysable properties regarding atomic states, emission of the line-spectra, etc., which are followed by two of the most foundational assumptions made by Bohr himself–

- The dynamical equilibrium of the systems in the stationary states can be discussed by help of the ordinary mechanics, while the passing of the systems between different stationary states cannot be treated on that basis.
- The latter process is followed by the emission of a homogeneous radiation, for which the relation between the frequency and the amount of energy emitted is the one given by Planck's theory.

The atomic model of Bohr was brought out with certain modifications in the Rutherford's model but then Bohr's model also ultimately came out with some peculiar limitations, for instance it could not explain the spectra obtained from larger atoms, it cannot predict the intensities of spectral lines etc.

Apart from atomic theory, there was another idea of Bohr which was challenged the father of quantum physics, Albert Einstein and this theory was brought out by three physicists - Niels Bohr, Hendrik Anthony Kramer and John Clarke Slater. The theory was entitled as Bohr–Kramers–Slater theory, which was a fundamental attempt to understand the interaction of matter and electromagnetic radiation. This theory wasn't fully established because it faced certain challenges in the experiment based on the statistical methods. Bohr had a huge objection towards Einstein's localized quanta of light, as according to him, any separate particle can't explain the interference effect of light and its wave properties. Another one of his well-known principles is named as Principle of Complementarity already explained in the initial phase disclosing the foundations to the quantum physics as a brief conceptualization. Some more examples of this principle are Time and Energy, photon, polarization, etc. In 1949, Bohr wrote an article titled "Discussions with Einstein on epistemological problems in atomic physics"; this is said to be the most profound exposure of the complementarity principle. After this principle, the Copenhagen principle is the commonly regarded as Bohr's one of the best interpretations on quantum mechanics which is also attributed to Werner Heisenberg. This agrees with the fact that quantum mechanics can only give probabilistic explanation for any system, or it also fundamentally says that a particle doesn't exist at a single position at any peculiar time but it exhibits superposition, which is mathematically expressible. With respect to the Copenhagen interpretation, the most fundamental underlying law is 'Born Rule', which gives rise to the probabilities in mathematical formalism that yields the measurement accordingly and hence the interpretation becomes more profound with its mathematics in various quantum aspects, for example wave-function collapse. This interpretation gives many fundamental explanations for many significant principles or thought experiments of quantum mechanics like double-slit diffraction, Schrödinger's cat thought experiment, etc. In terms oflarge quantum numbers, the Copenhagen interpretation refers the case to one of the other principles of Niels Bohr, the Correspondence principle, which suggests the reproduction of classical laws of physics in terms of modern quantum interpretations in order to generalize the large scale system including quantum numbers, energies, etc. After a certain time, Bohr's interpretation, which he had meant for quantum mechanics, turned into a metaphysical sense due to modern physicists like John Von-Neumann and a few more. The soul and the most rudimentary reason behind this was the phenomena of wave-function collapse because it was later demonstrated in such a way that made people think of the ontological analogy of quantum mechanics. Although there was a claim that Copenhagen interpretation didn't allow making such deviation from physics to metaphysics, but it is said that Einstein loosely took side of this transformation; although it is not rightly referred because Einstein wanted to understand the Universe in form of scientific and mathematical formalism, not in form of any metaphysical phenomenon. Anyway, the theories of Niels Bohr played a very significant role in various regimes and due to his splendid work and thus regarded as the third greatest thinker of quantum physics after Albert Einstein and Max Planck. The interpretations of Quantum Mechanics are mostly probabilistic, whereas the deterministic and local interpretation primarily includes the "Many-World Interpretation", given by Physicist Hugh Everett with his doctoral thesis in the year 1957 asserting that there is no wave function collapse and universal wave function refers objective reality. This became a vital and allied facet of decoherent principles further elucidated by Physicist Bryce Seligman DeWitt.

25

Werner Heisenberg – The Uncertainty Principle, Heisenberg Cut, Matrix Mechanics, Kramers-Heisenberg dispersion formula, Heisenberg's model for Statistical Mechanics & Heisenberg Group

What we observe is not nature itself, but nature exposed to our method of questioning. – Werner Heisenberg

Werner Heisenberg (1901-1976) was a German physicist best known for his famous Uncertainty Principle, which is one of the most fundamental principles of modern quantum mechanics and in some more advanced factors is also responsible for the rise of the concept of 'Quantum

Fluctuations', which justifies the provisional deviation in the energy proportion at particular point in space. He was awarded the Nobel Prize of Physics in the year 1932 "for the creation of quantum mechanics, the application of which has, inter alia, led to the discovery of the allotropic forms of hydrogen". Heisenberg also had a big hand in the Copenhagen interpretation along with Niels Bohr, as described previously. He also was one of those scientists who said to have favoured some sort metaphysical points in the consideration of quantum explanations; his perception over the Copenhagen interpretation leads an natural association with the wave-function collapse and this consequence is known as Heisenberg Cut, whose only major limitation is that it deals with only those principles, hypotheses or interpretations which has something vital to do with the collapse of wave-function, giving rise to the role of observer's information or conscious awareness for quantum events. Some refers that this hypothesis of Heisenberg deals with consciousness too which is a direct indication of observer effect as well, and this whole thing implies the application of this principle for the unification of quantum mechanics and metaphysics; however this cannot be determined for certain. Heisenberg is also one of the founders of Matrix Mechanics, which is a mathematical development of many fundamentals in terms of theoretical quantum measurements that is also concerned with the implications of some profound existing mathematical stuff to be in the account of quantum mechanics for example Fourier Transforms (it is a mathematical tool used to disintegrate some peculiar functions with certain properties to its inherent frequencies or wavelengths), Hilbert Spaces (they're a set of mathematical formulations which generalizes the Euclidean space from specific number of dimensions to arbitrary one or even infinite), etc. But this Matrix representation upset Einstein because it pointed towards the substituting elements of Newtonian conceptualization of space and time. Heisenberg also showed the consequence of electron as a cross-section for the photon scattering which most fundamentally uses the Correspondence Principle to the classical dispersion of light in mathematical ways and hence it becomes the Kramers–Heisenberg dispersion formula, formulized by Hans Kramers (or Hendrik Anthony Kramers) and Werner Heisenberg in 1925 and taken forward by Paul Dirac. This formula made a fundamental contribution to the explanation of Stimulated Emission in which a photon interacts with an atomic electron at its excited state with respect to photon's specific frequency. Apart from these, Heisenberg also contributed very well in the field of statistical mechanics for example - the notion of Heisenberg Model, which describes the transition between physical states (or phase transition) in terms of magnetic systems. The biggest deal in its description is the quantum spin of the magnetic systems. The pair of magnetic dipole moments as its natural repercussions comes out as an aftermath of one dimensional mathematical treatment in form of Hamiltonian (it is an operator which deals with the summation of kinetic and potential energies of all particles in any specific system) whereas the multi-polar facet of the model treated on the basis of a spate notion of spin-orbit reciprocity of that magnetic system which is called as the Multi-polar exchange interaction. His investigation, which to the led the extension of Langevin equation, works quite precisely for an open quantum system which represents the random evolution of quantum measurements as respective trajectories; and this formulation is also connected to one of his other contributions named as Heisenberg Picture, which has a fix open system with half of the time-dependent operators, and other entities like vectors that hold no dependency of time. Apart from physics, Heisenberg made some key contributions in mathematics as well, particularly concerned with abstract algebra like group theory, algebraic topology, etc. His ideas in that realm have made foundational contributions to the field of quantum mechanics, in its corresponding mathematical enhancements. For example, one of his noteworthy ideas in this area is Heisenberg Group, which is upper triangular matrix of specific form, and the two most basic outcomes of this is when the constituents of the matrix is integers then it will result in Discrete Heisenberg Group, and when the constituents will be real numbers then it will result in Continuous Heisenberg Group. The common thing among both of them is that they are nilpotent groups that occur in the principal linkage between field theory and group theory, which is better known as Galois Theory. I'm not explaining it deeply because it will surpass the abilities of general public to grasp but still, in simple language, one can understand the nilpotent groups as a workout of mathematical outcomes, which are constructed into some domains to terminate the previous or original function to extract the desirable solution from it; these domains are built up with certain notions like upper central series, lower central series, trivial subgroups etc. which play their respective roles in order to stabilize the finite lengths at different co-domains. Now, the key differences among both Discrete Heisenberg Group and Continuous Heisenberg Group is that one is fundamentally 'non-abelian group' (i.e. it has a compulsory availability of faintly one discrete pair of elements from that nilpotent or the intrinsic group) and the another one is 'lie group (which has organized form of continuous elements of the respective group) respectively. Heisenberg's perception of Matrix Mechanics gave rise to the new notion in the account of mathematical modelling of quantum mechanics, i.e. Banach Algebra, which are relative to the formulation of associative algebraic properties dealing both complex and real numbers simultaneously at a rudimentary level; especially the notion of "C^* - Algebra" which is particularly concerned with the Complex Hilbert Space as continuous linear operators and adjoints of topological norms or mappings.

26
Erwin Schrödinger – Schrödinger Equation, Affine field theory & Schrödinger Group

The scientist only imposes two things, namely truth and sincerity, imposes them upon himself and upon other scientists. -Erwin Schrödinger

Erwin Rudolf Josef Alexander Schrödinger or Erwin Schrödinger (1887-1961) was one of the most notable thinkers of the quantum mechanics, perhaps the fourth greatest after Albert Einstein, Max Planck and Niels Bohr. In 1933, he got the Nobel Prize of Physics shared with another pioneer physicist, Paul Adrien Maurice Dirac (or Paul Dirac) "for the discovery of new productive forms of atomic theory". He made some fundamental contributions in the field of quantum mechanics among which a few became quite famous, the first one is the Schrödinger Equation, which was postulated or formulated around the 1925. As it's known in quantum mechanics, particles exhibits wave-like characteristics and the Schrödinger Equation governs or shows the mathematical aspects of the behaviour of these waves by its description of the wave-function using a linear partial differential equation. The outcome of the equation contributes in many different areas of quantum mechanics like Unitary Principle (which reflects the time evolution of any certain quantum state), Copenhagen interpretation, T-Symmetry, etc. This equation in its native form is closely allied with the Hamilton-Jacobi Equation, which in itself a noteworthy subset of Hamilton-Jacobi-Bellman Equation that is a non-linear partial differential equation that is the underlying pathology for the optimal control theory referred to as the 'cost function', is useful in mapping event in the theory. Schrödinger Equation not only applies at the quantum level but also to the macroscopic level. It is also said that just after this equation, Paul Dirac found that both Schrödinger's and Heisenberg's versions of quantum mechanics were somehow identical; hence they both have same accountability regarding the ultimate version of quantum mechanics which takes both under its regime. Another great contribution of Schrödinger was his "Cat Experiment" which is mentioned several times in this book. It is said that even Schrödinger didn't like the probabilistic interpretation of quantum mechanics, thus he once even said,

"I don't like it, and I'm sorry I ever had anything to do with it".

Long after his early contributions, Schrödinger finally jumped into Einstein's legacy i.e. the Unified Field Theory, which suggests the unification of all fundamental forces of nature within the formulations of Einstein's Theory of General Relativity and in order to do that he came up with Schrödinger's Pure-Affine Theory or Affine Field Theory. In this theory, he intensively investigated the 'Affine Connections' as he thought the sole foundation to the differential geometry. Affine connections, fundamentally connects the neighbourhood tangent bundles of spaces over the smooth manifold as the core point. Although from this theory he made some good formulations like from the use of only affine connections, he successfully imposed the metric tensor through some mathematical establishments in the components of Riemannian Manifold, for e.g. Riemann—Christofell tensor or Riemann Curvature tensor, which represents the Riemannian Manifold in generalized form. But, he did not achieve his aim because mainly, Einstein was doubtful about this; hence Schrödinger left the idea and moved on some other points of physics.

Apart from quantum mechanics, he made some fundamental contributions in the field of mathematics which were similar to Heisenberg's, for example Schrödinger Group. The idea of Schrödinger Group has basically taken forward with the origin of his Schrödinger Equation only because it is a symmetry group of that equation as free particle. Even this notion of Schrödinger Group can be imposed on the Heisenberg Group with certain specific mathematical tools whose outcome are certain seminal changes in the Schrödinger Algebra which is a form of 'Lie Algebra' refers to that Group. Most profoundly, the Schrödinger Group can be merged with the Conformal Group (a transformational group of any topological space applies for the angle preservation to its own domain of space) in spatial dimensions if concerned with the relativistic aspect. A numerous applications are there of Schrödinger's Group or Schrödinger's Algebra, like it can be applied to a certain range in the Lie Algebra which possess a bilinear map with the co-domain of specific formulations.

Ultimately, anyone can easily understand how the ideas and theories of Schrödinger have successfully contributed to a lot of areas in theoretical physics and mathematics, which will have plenty of applications in future as well.

27
Arthur Holly Compton – Compton scattering & its Applications

My first feeling about the paper and the attitude is that it is absurd. - Arthur Holly Compton

Arthur Holly Compton (1892-1962) was an American physicist, famous for his 'Compton Effect' for which he was awarded the 1927 Nobel Prize of Physics jointly with Scottish physicist Charles Thomson Rees Wilson. Around the year 1923, Compton came up with his observation of the effect for which he is best known, i.e. the Compton Effect which starts with the photon scattering by a charged particle or specifically, electron. This phenomenon of scattering is called Compton scattering, which is a kind of inelastic scattering because the kinetic energy of the mutual incidence will not be conserved. The entire procedure of Compton scattering will lead to the increment of photon's wavelength and this is called the Compton Effect due to the change in the proportion of wavelength. The photon may be attributed as the emission of gamma-radiation or X-radiation. In the experimental analysis, electrons were treated as the free atomic entities thus each of them could contribute by the emission of certain reasonable secondary wavelets and the reason behind has to do with the energy of X-ray photon that was much higher as compared to electron's binding energy which led the entire experiment. This whole experiment and the notion of 'Compton Scattering' played a prominent role in disproving the entire notion of Bohr—Kramers—Slater theory, previously mentioned in the account of Niels Bohr. There is another notion of this scattering which takes place at the time when a charged particle passes on its energy to the photon, called as the 'Inverse Compton Scattering'. Also when the circular polarization of photons occurs with high energy magnetization then the process is known as 'Magnetic Compton Scattering' whose mathematical expression extensively deals with the electron momentum distribution in three dimensions referred to its spins. In the explanation of this scattering effect, Compton defined the 'Compton Wavelength', which is also a fundamental characteristic of a particle in the field of quantum mechanics. This notion of wavelength says that the Compton wavelength of a particle is equivalent to the wavelength of the corresponding photon whose energy is identical to the mass of that particle.

The Compton's scattering and other related phenomena are quite useful in various areas like in gamma-radiation spectroscopy, radiation therapy, etc. Also, the Inverse Compton scattering is very useful in cosmology and astrophysics with high energy physical systems like Cosmic Microwave Background radiation, when the photons are scattered to high energy electrons and then have a movement through hot gas adjoining by a cluster of galaxy. It can also be applied in the production or formation of the thermal spectrum from the surroundings of a black-hole, i.e. specifically at the center of our galaxy. The application of the scattering is also in the "X-ray Astronomy", which fundamentally deals with the observational detections regarding X-radiations.

Part 7:
A Brief Introduction to Some Enhanced Versions of Quantum Physics

28
Quantum Field Theory

God used beautiful mathematics in creating the world.
- Paul Dirac

After Einstein's Theory of General Relativity, quantum physics is the most precise and successful framework we have; and as general relativity has turned out in many other versions with the passage of time, the same thing has happened with quantum physics too. The investigation by plenty of brilliant minds in quantum mechanics has discovered its utility in several different areas of sciences including cosmology, particle physics, etc. The foundations of the quantum theories have been enhanced as the fundamentals of some other integrated facets of that. One of these facets is the Quantum Field Theory or QFT which is briefly defined in one of the initial chapters under the domain of Ontological Epistemology (more specifically in the point concerning quantum mysticism with neuroscience), where it is defined as a theoretical and mathematical framework which fundamentally deals with the unification of three of the finest entities of physics, i.e. classical field theory, Einstein's Theory of Special Relativity and quantum mechanics. But, most effectively, the core framework of the quantum field theory is defined as - each and every subatomic particle has a peculiar corresponding field, thus all the observed particles are merely localized oscillations or fluctuations in that respective field, i.e. surrounded regime of particles. In this way, the modern physics assumes that localized vibrations could be found everywhere because the entire space and time or throughout the universe these fields exists at all points, which satisfy all generally recognized or known subatomic particles. Hence, the most accurate way to conceive the subatomic realm is to relate those underlying fields of all the particles and then only, one can obtain profound results whether they're theoretical or mathematical. Apart from this, QFT is defined as the unification of classical field theory, Einstein's Theory of Special Relativity and quantum mechanics because when it comes to the mathematical representations with respect to any hypotheses, principle or theory regarding some new addition in the QFT, then just these three can fundamentally act; whereas the biggest and most notable limitation of QFT is that it excludes Einstein's Theory of General Relativity or the actual description of gravity, and thus resists the relative science to make the full description of nature. But despite this, QFT is still a very astonishing configuration because both the theoretical and mathematical descriptions developed by this version of quantum mechanics are most precise to the cardinal reality of nature; even brilliant physicist Richard Feynman contemplated the quantum field theory as "the jewel of physics". The whole account of QFT intensively contributes in various other constituents of physics; but its more deep-rooted as a primitive framework for explaining the physics of elementary particles and their corresponding interactions over the "fields", i.e. systems containing an infinite number of degrees of freedom, or contributes to the account of particle physics, in brief.

In particle physics, the QFT is exclusively applied in order to assemble the physical models of subatomic particles. Apart from particle physics, it is also applied in the study of physical behaviours of matter in both macroscopic and microscopic realms, i.e. condensed matter physics; QFT especially deals with multi-body system in the condensed matter physics, for e.g. quasi-particles (these have an excitation of multi-particle system with a dispersion relation similarly like the interaction of elementary particles). The mechanism of QFT in the condensed matter physics can be imagined with the help of Soviet physicist Lev Landau's thought on this, which indicates that the low-energy excitations in several condensed matter mechanisms could be defined pertaining to the interactions between group patterns of quasi-particles. The creation of quantum field theory is mainly attributed to those attempts subjected to relate the properties of quantum mechanics to the electromagnetic field; like de Broglie mentioned in his doctoral Thesis,

".....Based on an understanding of the relationship between frequency and energy, we proceed in this work from the assumption of existence of a certain periodic phenomenon of a yet to be determined character, which is to be attributed to each and every isolated energy parcel, and from the Planck-Einstein notion of proper mass, to a new theory. In addition, Relativity Theory requires that uniform motion of a material particle be associated with propagation of a certain wave for which the phase velocity is greater than that of light….."

But, mainly Paul Dirac is considered as the pioneer of QFT due to his constructed frameworks on the foundational aspects of QFT such as the notion of quantization of electromagnetic field and beyond. There is another notion as a subset of quantum field theory which is called as the 'Free Field Theory' or FFT, which deals with the aspects related to mass and motion regardless of the interactions of particles; a few are some noteworthy and fundamental frameworks which are identical in both quantum field theory and free field theory, one of such notions is the Gårding–Wightman axioms. These axioms were kind of mathematical trial initially focused on QFT by Swedish mathematician Lars Gårding and American physicist Arthur Strong Wightman. The axioms deals with a kind of Hilbert Spaces, which when imposed by the ten dimensional non-abelian lie group as the Minkowski's space-time or Poincaré group, i.e. instantaneous occurrence in the field is invariant, known as the 'Quantum Vacuum State' or the lowest possible energy state; giving rise to the unitarily. But, eventually its implications on QFT turned out to be on FFT also because the axioms don't take a lot of other interacting phenomena into their consideration, e.g. Gauge theories, effective field theory, etc. Although the effective field theory contributes in many dimensions of theoretical physics like general relativity, particle physics, statistical mechanics, statistical thermodynamics, condensed matter physics, etc. because it fundamentally uses the most effective set of degrees of freedom in order to explore or define the occurrence of physical phenomena at a specific scale of energy or just operates the radiation effects and dissipative systems. Similarly, the notion of Gauge Theory is related to the classical field theory that plays a significant role in the QFTs and other field theories in physics because they allow the Lagrangian to be invariant under certain local transformations by peculiar lie groups and thus synchronizes the derivatives of the degrees of freedom in the Lagrangian, which are almost like effective field theory.

The pioneers of the free field theory were three German physicists who also contributed a lot in the form of modern interpretations of quantum mechanics: Werner Heisenberg, Max Born and Ernst Pascual Jordan. Actually their intention was perhaps to define the terminology of something like QFT, because they fundamentally utilized the quantization of the selective classical field theories or what is called as the "Canonical Quantization" via intrinsic degrees of freedom of the field as an unbounded set of harmonic oscillators because if the condition is visualized as a one-dimensional vibrating string like a violin or guitar, string is displaced by some distance from its equilibrium position at any instance of time, which varies over time as the string fluctuates back and forth. Similarly, a harmonic oscillator is nothing but a mechanism of the classical mechanics which experiences a reinstate as force, i.e. proportional to the displacement when superseded from its equilibrium position. The analogy of one-dimensional string can be extended up to any number of dimensions for example, if the three-dimensional analysis in concerned, one can imagine a big hall - obviously filled with air with the air density consisting of an equilibrium valuation or merely the mean density as at every extremity in the hall; the sound wave causes air density to exhibit vibrations up to higher and lower values and hence the air density can be defined as a "field" because here the whole notion of air density has certain property with peculiar value at every point in the space of the hall with respect to time. But while determining FFT, they also supposed the absence of charge or current in the whole theory, hence such a phenomenon came out that had nothing to do with the 'interactive phenomena', but then also FFT by itself has some very fine applications in theoretical physics, which is quite significant.

Apart from all the aforementioned scientists, a few more contemporary scientists were there whose contributions successfully added certain significant and fundamental studies in QFT and one of them was the great physicist Enrico Fermi known as the "architect of nuclear age". He generally known for his contribution in statistical mechanics, the Fermi—Dirac Statistics which is attributed to both Enrico Fermi and Paul Dirac as they both discovered it independently; although it is said that Fermi was the first person to describe the phenomenon. This version of statistical mechanics precisely describes the distribution of peculiar indistinguishable particles with half-integer spin over similar kind of system which is thermodynamically stable or at equilibrium state. The entire conceptualization of Fermi-Dirac Statistics fundamentally follows the Pauli Exclusion Principle. These particles with the notion of half-integer spin are now called the "fermions", which is said to be coined by Paul Dirac, perhaps in the honour of Enrico Fermi. These fermions are the counterparts of 'bosons' originating from the Bose-Einstein Statistics that has just integer spin. The Fermi's contribution to the QFT is firstly the Fermions itself because as physicists later on efficiently discussed the whole system of Relativistic Quantum Field Theory or Relativistic Quantum Mechanics, they have become one of the core parts of interactions with some other elementary particles. Other than the consequences from the notion of Fermi-Dirac Statistics, Fermi's Interaction or Fermi Theory of beta-decay is also one of the early contributions to the area of QFT. The Fermi Theory of beta-decay basically describes the radioactive decay as intrinsic nuclide to an atom with unalike chemical elements with the same number of nucleons called isobar; of the emission from atomic nucleus of specific elements known as "beta" elements; this whole concept is known as "beta-decay". But this phenomenon is also has several direct analogies to QFT in which the interaction between four fermions fields at an identical point which exhibits the same occurrence as the space-time locality and due to this implication, the Fermi's interaction is sometimes also alternatively referred to as 'Four-fermions interaction'.

The one of the most key consequential phenomena of Fermi's interaction became a trailblazer for the 'Weak Nuclear Force', which is currently one of the crucial facets of many revolutionary ideas in Physics such as unified field theory, quantum field theory, etc. Due to this particular aftermath as weak nuclear force originated from the Four-fermions interaction, the entire theorization of Fermi became an approximation to the Effective Field Theory, specifically in low-energy formulations. The Fermi's terminology is certainly a kind of generalization of the Ambarzumian–Ivanenko hypothesis, which is attributed to two prominent physicists, Viktor Amazaspovich Ambartsumian and Dmitri Dmitrievich Ivanenko. The hypothesis foundationally explains the creation of massive particles. Ivanenko's another contribution is 'Ivanenko-Fock Coefficient', which also plays a significant role in the mathematical side of QFT because it defines the collateral displacement of 'spinors' in the curvature of space-time. Spinors are nothing but merely some forms of complex vector spaces, which have a fundamental property to be associated with the Euclidean Space as linear transformations. It can also formulate some foundational elements of Einstein's Theory of Special Relativity, i.e. the Lorentz transformations with respect to its combination with the notion of Minkowski Space, and as it's already known that the quantum field theory also includes the formulations of Special Relativity as it's foundational peculiarity; thus the Ivanenko-Fock Coefficient also gives rise to certain mathematical possibilities in various forms of QFT because it has several domains inside it, which means that the notion of QFT is not singular but a whole branch of theoretical physics that acts as a bridge between the quantum physics and particle physics and deals with various phenomena in its range of conceptualization, e.g. quantum electrodynamics, Relativistic Quantum Mechanics or Relativistic Quantum Field Theory and Topological Quantum Field Theory (TQFT), etc.

29
Quantum Electrodynamics & The Relativistic Quantum Mechanics

What I cannot create, I do not understand.
- Richard Phillips Feynman

Now we have successfully arrived at the point of discussion, which is considered as the firstly accounted quantum field theory. As in the last chapter, very fundamental aspects of QFT were introduced and it was also mentioned that the QFT is not a singular theory, but a collection of various different branches, which plays a remarkable role in an entirely distinguished and specific scenario or realm of both theoretical and experimental physics. English physicist Paul Adrien Maurice Dirac (or Paul Dirac) is regarded as the father of quantum field theory, although some other scientists of the same century, some physicists and mathematicians are also often credited for the some enhancements in the QFT, either in the form of their mathematical formulations or in the form of theoretical ideologies. These contemporary scientists mainly include three of the frontiers of quantum mechanics, i.e. Werner Heisenberg, Ernst Pascual Jordan, Max Born. But the core development or modernization of QFT took place with the great modern-age physicist and mathematician Richard Phillips Feynman, who is best known for his contributions in the area of quantum physics.

As defined in the previous chapter that quantum field theory is not a singular theory but a whole branch of physics which uses quantum methodologies to be imposed in the terms of particle physics. Hence, the first ever and the most accurate domain introduced in the realm of quantum field theory till date is the Quantum Electrodynamics or QED. As we know that the Classical Electrodynamics is fundamentally a kind of dealing with the interconnection between the currents and electric charges by using the mathematics and laws underlying the formulations of Newtonian physics; similarly, Quantum Electrodynamics deals with many such phenomena using the interaction between quantum entities, like light and matter. But unlike classical electrodynamics, QED deals with a lot of different aspects in a bigger regime as compared to the classical one because it deals with more rudimentary level of nature in the quantum scaling but that doesn't mean that the classical version of electromagnetism or electrodynamics is less effective because both have their separate prominence, applications and implications in different terms and methodologies. Even Richard Feynman, the modern founder of QFT or QED defined a few ideas that relate, even if slightly, the classical realm of electrodynamics to the modern interpretations of it. For example, the Wheeler–Feynman time-symmetric theory or Wheeler–Feynman absorber theory which is also attributed to the American physicist John Archibald Wheeler alongside Feynman. This theory is deciphering of electrodynamics based on the objective assumption that solutions to the electromagnetic field equations must act as invariant under the transformations of "T-symmetry". Although, the theory had a few limitations and drawbacks for instance, the Wheeler–Feynman absorber theory when considered with Mach's Principle suggests a disagreement with the self-interacting behaviour of elementary particles and this discards the notion of self-energies in QFT, although Feynman himself finally accepted that the self-interaction of elementary particles has to be there in order to conceive the theory more profoundly.

This was just a bit of extra information, so that we can now come back to Quantum Electrodynamics. So the notion of Quantum Electrodynamics or QED foundationally deals with one of the fields regarding the explanation of quantum field theories, i.e. the Electromagnetic Field or EM-field. The entire theoretical and mathematical delineations of QED only elaborate the descriptions or behaviour of EM-field; it also deals with the radical deportment of electron. The actual scenario starts with the limitations of the Schrödinger's equation because ultimately the equation is quite incompatible with Einstein's Theory of Special Relativity because of two underlying limitations; one is that it can't describe things moving anywhere near the speed of light and another that the equation supports the action of forces spontaneously at the same time, which is irrespective to the multiple frames of reference as it determines the evolution of particle's wave-function by fundamentally assuming only singular way of the passage of time, i.e. with respect to only one clock that is into the observer's frame of reference while on the other hand, Einstein's Relativity describes unification of space and time or both remain relatively sustain as constitutionally connected by the transformations of frames of reference; even this conceptualization of time and frames of reference can be considered as somehow one of the key reasons behind working of Schrödinger equation only for the slow moving objects because special theory of relativity has a very fundamental repercussion, which says that the passage of time depends on velocity. Schrödinger's Equation has one more drawback which restricts it to be congruous with the frameworks of QED itself and that is it's obstacle for the many particle systems, because the equation basically obeys the realm of position and momentum inclusively, the quantum state of every isolated particle and this whole system is ineffective due to the identity of particles which are obviously identical and exhibit mutual interactions in multiple ways thus the probability measurement eventually gets affected.

Hence, Paul Dirac around the 1928 came up with the first successful unification of quantum mechanics and Special Relativity with some very satisfactory methodologies and this formulation is famously known as the Dirac Equation. This equation has elegantly described the space-time evolution of a four-component spinor attributed to the behaviour carried as particle wave-function, which is denoted by the Greek letter – "psi", the equation also consisted of both Planck's constant and the speed of light, which indicates both quantum mechanics and special relativity respectively at the same time; and the whole mathematical terminology could explain and predict the motion of electrons to the speed of light and even in an electromagnetic field. Although a few contemporary scientists have sought such a theoretical construction with appropriate mathematics but have failed. Dirac's configuration has also been successful in getting over the limitations of Schrödinger's Equation from the account because he approached the theory in a very different way incorporating Einstein's Special Relativity; although Schrödinger himself was not much focused on the quantum field theory at that time, hence his core motive to define the equation was obviously not to unify quantum mechanics and Einstein's legacy. Dirac considered the quantization of electromagnetic field as a whole instead of quantizing physical factors of particles like position and momentum or following the alternating quantum states of each and every particle; and this was the birth of quantum electrodynamics and the relativistic quantum mechanics or relativistic quantum field theory, as the first foundational insight of QFT. Dirac fundamentally assumed each point in space as an infinite assemblage of simple harmonic oscillators similar to the case of photons, and can be illustrated as the minimum amplitude of quantum fluctuations; and this whole system turned out as nothing but a space itself of quantum states, including the physical properties like momentum and position like an array as defined previously regarding the simple harmonic oscillator. Thus, it can be said that Dirac included the physical notion but not in a simple direct way as in the Schrödinger's Equation; but with a key difference which makes the two separate interrelations and this is the distinction of mathematics, because the mathematical formulation of Dirac's Equation presumes an arrangement, whereas there is no data of specific assortment to the Schrödinger's mathematics, which can lead to a kind of consequential deviating phenomenon of probability quantification which is enough to give wrong predictions. Like Schrödinger's insight became a significant approach for the terminology of 'First Quantization' i.e. for a single-quantum system, Dirac is responsible for the 'Second Quantization' in the language of QFT which can generally called the Canonical Quantization. Second Quantization is simply for many-body system or to analyse the commuting number of quantum fluctuations or particles per state. The notion of Second Quantization is far more profound compared to the First Quantization and there are a number of reasons for this, for example the Second Quantization is applicable to all the tested elementary particles although there are certain underlying rules for each and every field corresponding to that elementary particle. The Second Quantization has quite efficiently extended several quantum mechanical laws to be applied in a bigger or more generalized regime like the extension of Pauli Exclusion Principle to fermions, etc. There is another aspect of particles which can't be explained by the Schrödinger's equation whereas, second quantization explains it in quite an accurate way; this is a whole process which plays a very noteworthy role in the quantum realm, i.e. the process of "creation and annihilation of particles". In the occurring phenomena of interaction of different or identical particles, they are created & destroyed all the time as an underlying procedure, for e.g. an electron can absorb or emit a photon. Even the fact is, quantum physics have a whole separate mathematical construction regarding this process of creation and annihilation, which are basically distributed in two forms named as creation operator and annihilation operator, or raising operator and lowering operator respectively; though both are accumulatively called as the Ladder operators. Each operator has a different and suitable mechanism for different particles, although both foundationally work on the eigenfunctions as concerned with the quantum harmonic oscillator. Sometimes, they are also used in mathematics, especially in the affine Lie algebra. The entire conceptualization of creation and annihilation in QFT soon became quite radical for the subatomic world. QED thus, came into the account of quantum physics with a lot of great and astonishing consequences in theoretical physics, specifically it was the first theory which has successfully performed the interaction of matter and radiation with an outstanding precision and the whole credit goes to Paul Dirac who beautifully unified two most elegant theories of the great Albert Einstein. But the story doesn't end here, in fact its a starting point of many bizarre ideas. One more side of Dirac still remains, which lead to the introduction of something which scientists never thought or realized before, this was the ideology of the "Dirac Sea". Schrödinger's equation had a problem with the notion of particle's inherent property of spin, especially when a magnetic field is present and thus the spin direction becomes quite significant; and as we already know the equation is not compatible with fast moving electrons and even electrons in an electromagnetic field, due to some sort of its intrinsically non-relativistic form. So, the scene again starts with the pioneer of QED or QFT, i.e. Paul Dirac when he was trying to obtain a relativistic version to the Schrödinger's equation in the same account of his famous Dirac Equation. He was trying to figure it out because he wanted the mathematics to work-out perfectly for electrons including their primordial properties. But when the basic formulations came into existence, he immediately realized that he needed a four-component spinor to extend the framework at the level at which he had thought, but wasn't known what really those two additional components indicated, because on a standard note, there are only two-component spinors required, for e.g. in the Pauli Exclusion Principle. Eventually, Dirac got his beautiful mathematical equation but this gap still remained to be filled, although it is said that he already had this foresight to see that it was going to turn in something very new due the mathematical elegance, which of course really matters in any theory. It's the great and remarkable history of how Paul Dirac made such contribution, offering solutions to many existing questions, but one major question remained in front of him regarding those extra two-components. Soon after the equation came out, an attempt was made in order to enumerate the energy of the electron via the imposition of this equation, and the result was even more bewildering and exciting because the aftermath of equation was allowing the existence of electrons in the states of negative energy too. But, this conceptualization of negative energy leads to some very abnormal phenomenon, that if considered deeply come out to be nearly impossible. Hence, Dirac came up with another idea, a theoretical model to resolve this which we call the "Dirac Sea". This concept can be understood by assuming an eternal deep sea that has its existence spread over the whole universe. The sea is of electrons which settled all of the energy states as negative with the range of from negative infinite, up to zero. Thus, accordingly the underlying condition comes out, as the only time of our interaction with an electron is when it consists of positive energy which would be at top of the sea due to our assumption of range and here we can bring in the Pauli Exclusion Principle, that if the energy state of this imagined sea is stable at a level, i.e. all are utterly full, then not even a single additional electron can be lost because the extremity of the presumed range lies at the top. The consequential projection that came out altogether from the notion of Dirac Sea is itself quite stunning and thoughtful because if one removes an electron from the surface of the sea, then it will leave a hole, which is supposed to be acted as an elementary particle and thus, the whole scenario in terms of that hole would seem to represent certain preternatural phenomena of traditional particles, for e.g. acting like it has the mass of that absent electron, behaving like holding positive charge, i.e. opposite to that of the charge of electron, or consisting the corresponding inertia etc. The association of all these properties indicates a very significant notion of physics, i.e. the formation of Anti-matter. Although, the imagination of the Dirac Sea doesn't actually exist, but it certainly defines a very significant and real aspect underlying the description of QED and all kinds of quantum field theories, i.e. the idea of Quantum Field; and as we now know that all the elementary particles have a respective compatible "field" and in those fields both facets of the matter and anti-matter exists because the comprehension of Dirac's negative energy solutions to the equation defines the existence of anti-matter not any kind of real holes and these anti-matters are nothing but the fluctuating constituent of the same quantum field that are exhibited by its counterpart. Initially, it was said that Dirac through that anti-particle conclusion identified proton due to opposite charge although almost after one year, he changed his mind and postulated the positron instead of the counterpart of proton. He also determined a reaction which showed the formation of two photons by annihilation of electron and positron. Similarly, when a particle and its counterpart i.e. anti-particle collide, it eventually causes the annihilation process and as attributed to Einstein's mass-energy equivalence, both forms and manifestations of particles exhibits transformation to radiation. Dirac's conclusion of anti-matter raised certain questions, but after a very positive involvement of a few contemporary scientists, like Julius Robert Oppenheimer, Igor Yevgenyevich Tamm, etc. Even the great physicist and mathematician Hermann Klaus Hugo Weyl played a significant role in it by concluding the symmetric property of negative and positive charges in the Dirac Sea, specifically the 'hole theory' and this conclusion resolved the problem of the distinction between masses of electron and proton, which Dirac himself visualized as an aftermath to the electromagnetic interactions with respect to the imagined notion of Dirac Sea.

Soon after this astonishing postulation of positron or anti-electron, its existence was experimentally confirmed in the fullness of time by American physicist Carl David Anderson during his investigations on the cosmic rays collisions via a particle detector named "Wilson Cloud Chamber", which is basically utilized in order to analyse the transition of ionizing radiation and is proceeded by containing a systematic and suitable milieu of supersaturated addition of alcohol and water in the vapour form. Anderson won Nobel Prize for Physics in 1936 for his discovery of the positron, which he shared with another well-known physicist Victor Franz Hess who won it for the discovery of cosmic radiation. Thus, the mystery behind those four-component electrons was finally resolved with a very useful solution which is now used in various aspects of theoretical physics.

Apart from all of these, the existence of another counterpart of one of the elementary particles was successfully confirmed by Emilio Gino Segrè and Owen Chamberlain, i.e. of anti-proton; both the scientists were awarded Nobel Prize for Physics for their discovery in 1959. With the passage of time, a lot of advancement took place in this field due to one single contribution by the legendary Paul Dirac. At the moment, we have the anti-matter counterpart of hydrogen, i.e. anti-hydrogen, which is basically made up of positron and anti-proton and this whole aspect of hydrogen consists of some very novel characteristics such as, it is postulated that perhaps anti-hydrogen exhibits the gravitational interaction of anti-matter and similar to the natural hydrogen atom; and although it's a very fascinating point to research, this whole terminology is yet to observed or measured under a particular methodology by scientists via technological aspects.

This was just a beginning of quantum electrodynamics or the relativistic quantum mechanics because the core development in these areas is attributed to three most eminent modern scientists, i.e. Richard Feynman, Julian Schwinger and Sin-Itiro Tomonaga. Although, there is another brilliant modern physicist who has contributed widely to an aspect of fine tuned mathematical ideas in the modernization of QED, Freeman Dyson.

But, among all of them, one man especially stands out with his extraordinary ideas, which are intensively responsible in the development of the quantum field theory to a different level with some esteemed and noteworthy enhancements, and this scientist is none other than Richard Feynman. Feynman was an enormously curious person since the early days of his life and the consequence of his curiosity turned out as a historical quantum masterpiece. As we know Thomas Young's experiment has two slits and the theoretical perceptions as physical properties concerning an aftermath to that experiment has fundamental involvement of those slits in order to measure or predict some relevant aspects, for e.g. the probability amplitude formulated by the wave-particle duality relation along with its relationship to quantum optics, also makes the Bohr's Complementarity principle more profound in mathematical terms and utilizes the formulations of Max Born. But if one assumes something that seems to be absurd but is actually not due to its intrinsic rationality, the assumption is – "what kind of mathematical representations will be devised if there are infinite slits?" Well, the biggest problem ahead of this question is the number of paths leading up to the resulting interference, but one has to realize that it's very important, because especially at the quantum level things must be considered with a certain level of sensitivity due to a very microscopic framework that governs the whole synopsis. In the mainstream notion of just quantum mechanics, perhaps we don't have to consider this assumption of 'infinite system' to great extent due to the lack of intense interactive phenomena; but if the area of quantum field theories is concerned then it's a very important point of analysing the true and precise finite probabilities, and not just the possibilities but beyond predictions or calculations in mathematical expressions because the QFTs are all about interactions in very different quantum fields. The reason behind this assumption is simple, because there are infinity ways to go from one point to another, if we don't have an idea of all these paths then we would end-up with inaccurate predictions. Now the question is can we know about all these infinite paths explicitly, the answer is certainly no! Because we are dealing with the quantum world which is intrinsically non-deterministic, we can't determine the paths with exact certainty. But there is a separate regulative aspect to this problem for we can know it's very precise working mathematics, which can give us finite evaluation of infinite paths, including those which are very tough to think with respect to the fundamentals of arbitrary points in space-time. The mathematics we are talking about is attributed to Richard Feynman as Feynman's Path Integral Formulation. Feynman's technique for the path integral formulations is one of the most foundational mathematical framework ever constructed which explicitly allows us to trace all the possible paths that can be recognized as reaches or interact from one point to another, even those paths which are quite tough to imagine or conceive as mentioned above. Although the formulation obviously doesn't obey the deterministic nature of universe, or Einstein's perceptions, because the dealing is with the subatomic level but still it's not about exactly the superposition principle despite the fact that to some extent it is due to the native quantum property. But, quantum electrodynamics or some other quantum field theories are one of the finest and stunningly deterministic theories in quantum physics apart from Einstein's legacies of quantum and statistical mechanics; especially the rudimentary account of Feynman's path integrals are very deterministic as compared to other quantum mechanical formulations because they are connected to the classical theories up to some extent. The terminology of Feynman's work regarding this notion of infinite paths fundamentally takes advantage of the notion of "Principle of Least action" in the quantum realm, with the foundational imposition of certain quantum mechanical aspects regarding its native mathematical abstracts.

The most rudimentary idea on the way of this formulation of path integrals is the notion of 'time' to be taken as slices, or in order to consider the entire journey from one point to another as all possible paths, time would be generalized into small distinctive intervals, but manifestly originating from an absolutely identical aspect of physical laws. These slices in the form of time-intervals give rise to the tracing of all peculiar paths attributed to particles in the space with certain patterns. But one can easily imagine that this space-time analogy as described cannot handle or formulate the whole scenario because there arise some very fundamental problems; for example, it's quite natural that certain paths from this system of manifestation can be worked out in order to decode utilizing the domain including laws of physics, but there is an even spacious likelihood that a lot of paths can be involved in that procedure which have nothing to do with the rationality of physics, some even violate very primitive realities of nature or physics, specifically those of quantum world. He then included a very elegant principle from the world of classical physics just mentioned previously, the Principle of Least action or Principle of Stationary action which is basically a variational principle stating that an object will perpetually follow that path which reduces the notion of 'action'. As additional information, one should know the methodology of 'Action'; in physics, this term is certainly referred to the commensurable as relocation of both kinetic and potential energy or Lagrangian over a path with respect to the distinction of time so that it can generalize over the respective trajectories of different systems as their dynamics. The principle of least action is worked-out in multiple directions of theoretical physics and mathematics, for example in Hamiltonian and Newtonian mechanics, etc. The principle also has some very outstanding significance in terms of the Relativity Theory; the whole aspect of this principle gives a very meticulous explanation of the measurement of time in certain trajectory of any object over the curvature of space-time in cosmos and this is determined in two very deep-rooted convictions, i.e. "proper time" and "world line"; both of these are interconnected to each other because "world line" is attributed to the object as it's path i.e. over the four-dimensional space-time and if this concept is extended up to the fabrication of 'time', then it becomes the notion called 'proper time'. It plays a very vital role, especially in Einstein's Theory of General Relativity by allowing it to yield Einstein Field Equations via this principle in terms of proper sign-conventions which makes the formulation very effective and this generalization is known as the Einstein—Hilbert Action; apart from this, when the classical ideology of the large-scale universe is considered then reducing the 'proper time' allows to determine all the equations of motion. The principle was in some sense specifically extended in the term of classical realm by two legendary physicists and mathematicians of classical times, Joseph-Louis Lagrange and Leonhard Euler respectively with the famous Euler—Lagrange Equation, which is a second order partial differential equation that exhibit functional solutions with respect to certain kind of stationary functions; whereas the modern perceptions were extended by great scientist Paul Dirac and after him, it was further formulated by two of the most prominent founders of modern interpretations or advancement of quantum field theories, by American physicists and mathematicians Richard Feynman and Julian Seymour Schwinger; even as we know this whole consideration goes towards the path integrals given by Feynman.

The whole principle became one of the most fundamental keys as a kind of universal ideology, that objects always approach the path in terms of least action; although there is undoubtedly not any kind of 'single path' aspect in the subatomic or quantum world. Feynman's path integrals doesn't deals with the occurrence of individual probabilities in a system, i.e. for the number of paths, but each separate path play its own peculiar role for the notion of probability amplitude and this undoubtedly is a fundamental requirement to any quantum system. Feynman's approach is more powerful as compared to a lot of other quantum mechanical formulations, perhaps because of the classical touch in it, which is certainly a function of the quantum scaled path as measurement through the entire space-time and thus, the paths from any one point to another one in the large-scale universe turns out as a rational path of those infinite paths so far. Despite the analogy of space-time in these paths which treats both symmetrically, it is instinctively compatible with Einstein's Special Relativity and this makes the path integrals naturally consistent with the quantum field theory or electrodynamics, because Paul Dirac has also fundamentally come up with an elegant unification of quantum mechanics with special relativity.

Feynman's formulation is sometimes also known as the "sum over histories" because instead of summing up all attainable paths by all particles, it takes all possible histories of quantum fields into account, i.e. motion and energy-amplitude of any photon can be justified with their discrete deviation of the corresponding excitations in an electromagnetic field as it basically considers universe as oscillating fields such that it takes over the motion of particles. The path integral formulation is also a very significant tool in statistical mechanics which mainly arise from the notion of 'Wick Rotation', which give rise to the concept of Euclidean Path Integrals because it leads to a conversion of Lorentzian to Euclidean in the geometry of space-time or specifically, the terminology of Minkowski space. This concept played one of the key roles in order to link the notion of path integrals and statistical mechanics in form of various other vital interpretations like the Feynman–Kac formula, which fundamentally connects the aspects of stochastic processes and parabolic partial differential equations. The path integrals turned out to be one of the most precise mathematical frameworks in the account of quantum electrodynamics which has a lot of applications in various regimes of quantum physics; but Feynman had another insight into QED which deserves mention otherwise one aspect will remain unfilled up to the end, and that contribution is Feynman Diagrams. There are a plenty of complex aspects to which the Feynman's path integrals can deal, but they were not generalizations to those aspects in mathematical terms, these were phenomena in themselves. For example a photon is travelling between two distinctive points forming a certain path that instantaneously becomes an implicit electron-positron pair just before its annihilation into genuine photon at the end point of destination or path, and even more such cases can be there in terms of electron emission or absorption and beyond in much complex interactions; in order to have some generalized insights, Richard Feynman, around the year 1948 introduced a set of figures which are now called Feynman Diagrams. These diagrams deal with tons of possibilities, but there are certain set of rules in this concept which make it quite precise for measurements in QED or QFTs. There are some very fascinating techniques or characteristics in this elegant methodology but one of the most significant is that the diagrams are not merely an explanation of interaction or processes but corresponding to a peculiar mathematical representation or equations which are due to some key generalizations from these diagrams. Once the equations arise out from the respective diagrams, they are then treated with some mathematical formulations, for e.g. the Feynman's path integrals, etc. But sometimes the mathematics comes out to be in very complex form, i.e. quite difficult to resolve it in order to predict or calculate some required abstracts, but here we have a mathematical solution to this problem of complex consequence and the solution is Perturbation Theory, which suggests mathematical alternatives to an obtained equation, from Feynman's diagrams too. It says that one can replace the originally formed complex equation with another equation as an approximation, which can be easily resolved and lead the mathematical structure to predict and make quantum measurements. Perturbation theory is applied in several different conditions apart from the diagrams, and it's sometimes referred by some other names also due to the quantum constrains, for example when it determines the 'weak perturbation', which express the rate of transition from an isolated quantum system to a set of quantum systems in terms of energy eigenstates; then it is called "Fermi's golden rule", although the idea of perturbation is again attributed to Paul Dirac. On the other hand, techniques in the perturbation theory, sometime require expansion to make solutions more productive, thus there are a few settled methods to that in QFTs which includes some specific factors that plays foremost roles in synthesizing mathematics with elegant ways; for example in order to extract the derivatives of particular energy states, Hellmann–Feynman theorem is key whereas, the perception of Time-dependent perturbation theory can be supported via the notion of Dyson Series which defines the time-evolution similar to Schrödinger's equation but with different techniques. Although; contributions of both Schrödinger and Heisenberg certainly give a prime contribution for formalizing the expansion i.e. the intermediate state of 'Schrödinger picture' and 'Heisenberg picture' or 'Dirac Picture' in combined form, plays a dominant role because they both deal with the time-evolution in different underlying forms or the distinction holds in fundamental operators which are with respect to states; and hence the resulting "picture" from both i.e. "Dirac Picture" is an especial prototype of unitary transformations while imposed in certain unique quantum mechanical phenomena.

It is also said that Schwinger, with his another contemporary physicist Sin-Itiro Tomonaga conceived all the three pictures in such a way that they exhibit the covariance of perturbation theory. But, similarly Schrödinger's formulation, when taken in another form gives fundamental idea for the Time-independent perturbation theory imposing a self-adjoint operator, i.e. "Hermitian" dealing with orthogonal keys to finite-dimensional complex vector space. Perturbation theory also uses various mathematical tools of topology, differential geometry and beyond to generate proper mathematics; it also utilizes the S-Matrix or Scattering Matrix, which itself is a mathematical framework in Quantum Field Theory and elaborates certain very elegant concepts in QFT like field theory, which possess invariance under the Conformal transformations i.e. the Conformal Field Theory or CFT. Sometimes, interactions and fabrication of particles can be in the form of loop in any intermediate state and this gives rise to another problem regarding mathematical complexities like enormous dominance infinities in the resulted equation and many more, then even perturbation theory too shows its limitations and at such a moment the entire concept of renormalisation takes place which again consists of certain set of rules which helps the equations to get rid of those mathematical obstacles.

Although, there are two very famous notions which also treat these difficulties but they're not the substitutions of renormalization; the two formulations are Feynman parametrization and Schwinger parametrization named after Richard Feynman and Julian Schwinger respectively. The most common aspect of both the perturbation theory and renormalization is that it underlies the Lagrangian in their respective formulations. There are a few things that really matter in Feynman diagrams, but the most amazing among all of the fundamental parameters is the direction of time. Let's say there is an interaction with the participation of electron, its anti-matter counterpart or positron and photon; then we know that electron always exhibits forward in time if plotted in any space-time graph and the photon will have a wavy behaviour in that graph but what about the positron. So, great physicist and mathematician Ernst Stueckelberg proposed that a positron can travel backwards in time and the idea also convinced Richard Feynman, thus there is a very special role of the direction of time in these diagrams. Hence this perception is now called the Feynman–Stueckelberg interpretation. Apart from this, Ernst Stueckelberg contributed a lot in QFT also, for example he successfully determined a new and very important discovery in the perturbation theory about the covariance of Lorentzian. Like path integrals, Feynman diagrams can also be with multiple forms of figures for an identical interaction but the due to strong mathematical account in its procedure, one is allowed to predict or calculate most precise probability amplitude in the whole system of diagrams. After Richard Feynman, Julian Schwinger also played a very significant role in the development of QFT and Quantum Electrodynamics as he intensively worked in many aspects of these. One of his notable ideas includes the investigation on famous Green's function (named after great mathematician George Greene); the notion belongs to inhomogeneous linear differential operator under the domain of boundary conditions that also includes the superposition and Dirac delta function which is a mathematical abstract with plenty of applications such as it used to generalize the distribution of physical entities so that it applies to statistical mechanics too, whereas the Green's function applied in QFT or QED when the many-body system is concerned because at that point, the function fundamentally acts like the correlation function, thus it also play a vital role in other fields of physics or quantum physics like aerodynamics, statistical field theory, etc. Green's function is also applicable in condensed matter physics, for example the Landau–Fermi liquid theory, which primarily deals with the interaction of fermions that explains the metals at low temperature in the standard condition; although this idea breaks down for one-dimensional analysis, hence one of the pioneer of QED, i.e. Sin-Itiro Tomonaga defined his model which was about the interaction of electrons in a one-dimensional conductor. Schwinger's contribution in QFT from Green's function is referred as kind of equations of motion corresponding to that function, these set of functional equations are called as Schwinger–Dyson equations attributed to both Julian Schwinger and Freeman Dyson; another specific solution to these equations were defined in the two-particle relativistic systems that exhibits bound states and covariance which is constitutionally used in various branches in the account of theoretical physics for e.g. relativistic quantum mechanics, high energy physics, etc. This two-particle extracted form is now known as the Bethe–Salpeter equation, which is named after two very prominent and contemporary physicists of Feynman - Hans Albrecht Bethe and Edwin Ernest Salpeter. Schwinger also developed a variational orientation to QFT that is called the 'Schwinger's quantum action principle' and it contrasts from the classical interpretation used in the Feynman's path integrals as it has its own boundary conditions that formulate the Schwinger's principle mathematically. Schwinger's insight also played quite an effective role in the process of creation of electron-positron pair as spur-of-the-moment in strong electric field which was the enhancement of the ideas of physicist Fritz Eduard Josef Maria Sauter, whose work was extended by Heisenberg and Hans Euler; Schwinger's advancement is this study is also known as "Schwinger mechanism" or "Schwinger Effect".

Julian Schwinger got the Nobel Prize of Physics in 1965, which he shared with two other pioneers of quantum theory, Sin-Itiro Tomonaga and Richard Phillips Feynman for their "fundamental work in quantum electrodynamics, with deep-ploughing consequences for the physics of elementary particles", which they undoubtedly deserved for contributing towards the sciences and humanity with such a precise theory.

Another pioneer of QED, Sin-Itiro Tomonaga's work is not generally known but it is said that he and Schwinger shared almost similar kind of insights on the subject. Although both Schwinger and Feynman were seeking almost same aspects of field theory at a very microscopic level in terms of these interactions, vibrations of fields, etc, somehow their approach was quite distinctive because as we know that Feynman's perspective was creative in form of pictures, tricks and other fascinating ways whereas, on the other hand Schwinger's was more about the mathematical constructions of certain systems. But in the fullness of time, the results were almost same, though sometimes in different directions to each other. But everyone has their own natural style to obtain reality so there is nothing wrong in anyone's unique perception; there is merely a distinction of languages or creativity in order to seek true aspects of universal searches because creativity has no boundaries in any realm.

30
Quantum Field Theories for Strong & Weak Interactions

Think how hard physics would be if particles could think.
- Murray Gell-Mann

We have now discussed some very significant and foundational results of quantum field theories in different aspects, but quantum electrodynamics is only a description or mechanism of just one force, i.e. the electromagnetic field. But when the matter comes to the mechanism of weak and strong interactions deep inside an atom, we have some other concepts that are enormously helpful and worked out in just last few decades in the quantum world. In the account of QED, the notion of Dirac Equation has already been introduced, but apart from that there is another very significant formulation which is basically a relativistic version of Schrödinger's equation i.e. the Fock-Klein-Gorden Equation or Klein-Gorden Equation, which is named after three very prominent scientists - Oskar Benjamin Klein, Walter Gordon and Vladimir Aleksandrovich Fock. But certainly it's not as profound as Dirac's formulation because it has certain limitations and it's not very predictive about the nature of reality like counterparts of elementary particles or anti-matter; but nevertheless it has its own applications in the regime of specific particles like bosons. The equation is mathematically very informative as it deals with covariant Lorentzian as second-order into the Scalar electrodynamics or with the symmetry of U(1) as gauge field. The scalar field is itself a very significant conceptualization for various aspects of theoretical physics and mathematics; it also contributes to Higgs mechanism or Higgs bosons regarding the symmetry. Dirac equation is essential for electrons whereas, the Klein—Gorden Equation is for the spin-less particles like 'pions', which are the lightest forms of hadronic particles called 'Mesons' that are the composition of one quark and one anti-quark and these are bounded by strong force because they are implicitly minute and work-out intrinsically for the other subatomic particles or constituents of an atomic nuclei known as nucleons, i.e. protons and neutrons. Quarks exhibits SU(3) symmetry and were theorized by physicist, Murray Gell-Mann and also it is said that certain scientists came up with similar thoughts independently, around 1964. These symmetry groups are quite technical aspects so perhaps I can't make general readers understand with much depth, but let's try superficially to make them simple and generally conceivable. First, one has to know what exactly 'symmetry' is in the field of physics. Symmetry is basically a property of certain theory or its corresponding mathematical structure, which remains invariant under peculiar transformations respective of some specific underlying aspects. One can also understand this concept of symmetry with a sphere - if a transparent sphere is there with no additional appearance on it, even if the sphere is rotated it will not be realized due to its symmetric nature. Symmetry plays a very crucial role in a lot of facets of physics like in quantum field theories. U(p) is a definite form of an array which is commonly known as the Unitary Group and this group has a special case that is known as the Special Unitary Group that is represented by SU(p), which leads to the resulting scalar value as "1" to that unitary group, called 'determinant' and 'p' is any specific degree which an essential part because with respect to this notion of degree only the unitary groups are mathematically evaluated. These unitary groups are basically the lie group in "real" form and this is some sort of mathematical tool to represent certain specific aspects in mathematics and theoretical physics, including the symmetry of any particular suitable group of formulation.

So, coming back to the notion of quarks which are fundamentally of six types and are noted as - up, down, strange, charm, bottom and top.

As photons are necessary in QED, similarly for the field theory of strong interactive forces, another mass-less entities are there which are called 'gluons' and are subatomic particles that work in the transmission of bounding the strong force of quarks in the fabrication of hadrons which are the composition of two or more quarks. Gluons are of eight types whose combinations are based on the notion of Colour Charge, manifested as Red, Green and Blue; although they have nothing to do with the ordinary colours. If one goes with the combinations only according to colour charges and their counterparts, then it will end up with nine different arrays, but the respective concept of their gauge symmetries will also work because they are basically some sort of 'Bosonic' in the vector form as any gauge field.

There is a whole theory regarding the study of strong interactions which is called Quantum Chromodynamics or in short, QCD. QCD is also a kind of quantum field theory, but it has its own specialities in this regime like other QFTs. It should be known that intrinsically, quantum electrodynamics exhibits the symmetry group of U(1), similarly QCD has the symmetry group of SU(3), in fact both of the QFTs fundamentally include the notion of non-abelian gauge theory in their respective groups. Similarly like colour charges, the strong force is sometimes also called 'Colour Force'.

QED and QCD both have some similarities and some differences, but they both can be easily understood if related to each other like a few differences have already been explained previously in this book. Although, QCD came later and has evolved in last few decades, we have developed plenty of mathematics to elaborate it with more effectiveness and precision; for example, the interaction of gluons between quarks are described by 'Gluon field strength tensor' or if taken in more generalized form, then its Gluon Field. QCD uses many underlying and startling techniques to resolve the mathematical and even experimental obstacles, for e.g. the Perturbation theory or Chiral Perturbation theory which is formulated on the fundamentals of the "Effective Field Theory" using certain symmetries like the notion of Chiral Symmetry, which restricts any transformation or rotation to form a peculiar mirror image. As we know from one of the initial chapters that effective field theories fundamentally demands the Lagrangian coherence; but in the low-energy version of QCD which becomes the non-perturbative; Lagrangian in terms of quarks and gluons replaced by hadrons because in QCD, the colour charges can't be firm in isolated form which is known as 'Colour Confinement'; the low-energy QCD can be determined using Partition Function, which plays a key role in generalizing the quantum mechanical correlation functions. Lattice QCD is also another technique specifically concerned with the non-perturbative aspect of quantum chromodynamics, because the Lattice QCD is a Lattice gauge theory, which defines the space-time as any discrete fabrication of lattice form, hence there are also some very particular terminologies in this theory or methodology; for e.g. the 'Monte Carlo methods' has the consequential approach as computational algorithms hence they are also used in optimizations, statistical field theory, etc. Certain extracted forms of the Monte Carlo methods which determine the action of uncertainties as any modelling of risk analysis, sometimes also referred as the Monte Carlo Simulations, which utilizes some very fundamental tools of theoretical physics or mathematics like gauge assortments, wick rotation, etc. to formulate the abstracts, this technique simulation contributes not only in quantum regime but also in various different and profound areas of studies like quantitative analysis in financial mathematics, probability theory, etc. Up to the current time, the notion of Lattice QCD basically applies at the low physical domains of predictions and calculations; hence there are no restrictions over the numerical sign conventions while deducing the mathematics from current techniques introduced into the consideration of QCD. The non-perturbative nature in terms of Lattice or Lattice gauge theory in quantum and particle physics, while imposing the renormalization to certain techniques of Lattice QCD, leads to an approximate conversion to perturbation theory in forms of Lattices only, and again it has it's some very own and foundational methods and quantum approaches to define several significant results in QCD and even some other parts of physics too like condensed matter physics. The techniques involved in it are some of those which are quite rudimentary to the Quantum Field Theories or Quantum Physics, for e.g. the S-Matrix, QCD sum rule, etc. S-matrix or Scattering matrix has already been used a few times previously but the QCD sum rule is a fundamental to apply perturbation in QCD in some very distinct cases via the gauge invariants in the form of an axiom called the 'operator product expansion', which is developed under the domain of QFTs to synthesize the product of certain fields as the sum over those fields in a generalized way.

Up till now, we have discussed quantum chromodynamics at low-physical ranges or in asymptotic degrees of freedom with the help some astonishing concepts, but at high-physical domains like high density, temperature etc.; we have some defined methods to deal with. So, lets start with the concept known as QCD matter or sometimes also called the Quark matter, which deals with the corresponding degrees of freedom with respect to both quark and gluons in any uniform transitional space at higher states; although quarks can also be taken as a fundamental degree of freedom in the quark matter. Composite subatomic particles called baryons also play a noteworthy role in higher physical states of QCD, which are classified as fermions due to the spin and comes under the hadrons that exhibits odd number of valence quarks, i.e. minimum three. At specific range of temperature with adequately high density of baryons, which tends to at least above the nuclear one, then the high physical states in quark matter has a kind of prediction regarding the imposition of Colour Superconductivity, which is significantly known for this occurrence only. When it comes to statistical implementations of strong interactions, then there are a few respective convictions which initializes with the Quark-gluon plasma (QGP), which tends to the thermal equilibrium irrespective to the colour charges and underlies to the elements of Stefan—Boltzmann analogy under relativistic conditions and even in the terms of Quantum Hydrodynamics, which also have certain very vital contributions in condensed matter physics; apart from this, there are a few underlying conditions which are considerable in QGP, for example the temperature must be higher than that point where hadrons attains instability. i.e. Hagedorn temperature. Apart from this, the QGP is also predicted as one of the initial consequences of the big-bang. At the point of study in high densities, the theory of Fermi liquid also plays a vital role as it takes the QCD matter into its domain. The Fermi Liquid Theory is an abstract model of particle physics which basically deals with interaction of fermions but gives many rudimentary results at certain low temperatures with respect to any standard critical temperature. For the strong interactions, there is a very significant subset to the QCD, which deals with the whole scenario by performing a constructs of relativistic many-body system regarding baryons and mesons, thus we have another theory which is known as Quantum Hadrodynamics (QHD). QHD more related to the nuclear aspects and this is basically a version of Effective field theory only, which also leads to the precise analysis of something known as hadron-hadron interactions.

Such that there are a plenty of ideas developed in quantum chromodynamics or to describe the strong interactions through the concept of quantum fields. Similarly, we have another force which is definable under the perspectives and formulations with QFTs and that force is Electroweak Force, and thus the interaction is obviously known as Electroweak Interaction. Electroweak interaction specifically includes the types of bosons categorized as 'W & Z bosons', which are intermediate bosons manifested as vectors. The quantum field theory that deals with electroweak interactions is generally known by the same name only, but still the more precise terminology is Quantum Flavor Dynamics (QFD). Most of the initial aspects regarding electroweak interaction are attributed to three very prominent scientists, Sheldon lee Glashow, Mohammad Abdus Salam and Steven Weinberg; although it is said that Glashow conceived the idea at the very first stage. Although, there are certain identical regimes that are applicable to both QFD and some other QFTs, even the theory is developed at the level where we have the unification of two forms of forces among the four fundamentals, i.e. electromagnetism and weak interaction between elementary particles; also all the three pioneer physicists of electroweak theory awarded the Nobel Prize of Physics in 1979 for their contributions to the theory of "the unified weak and electromagnetic interaction between elementary particles, including, inter alia, the prediction of the weak neutral current." This idea of unification of weak and electromagnetic interactions is specifically known as Wienberg—Salam theory; also the idea is elegantly expressed with certain gauge group along with some more peculiarities of mathematics and like QCD, which exhibits its own gauge group in an indistinguishable way, this confederation or analogy is also mathematically described as the conception of Yang—Mills theory theorized by two scientists named Chen Ning Yang and Robert Laurence Mills.

Apart from these facts, theoretically, there are some reasons that led this possibility of unification. For example, one of the most radical reasons is the notion of spontaneous symmetry breaking that applies to both 'W & Z bosons' and photons. The spontaneous symmetry breaking can be understood in two different ways; theoretically it's the spontaneous breakdown of intrinsic symmetry and the ultimate result is the asymmetric state; and on the other hand mathematically it can be understood by the distinction of symmetries in the terms of Lagrangian field theory and vacuum solutions of Einstein's Field Equations, which leads to the vanishing of Stress-Energy-Momentum tensor - a generalization of corresponding tensor in Newtonian physics used to define the density or flux of energy and momentum over space-time. Additionally, one should be familiar that 'tensor' is a mathematical tool with generalizes the respective properties abstract objects over certain perception of 'space'. We can understand the concept of spontaneous symmetry breaking by the Higgs mechanism too because there are some symmetry groups where the inclination of mass for gauge bosons violates the symmetry and the same thing happened to the well-known 'Higgs boson' which has a respective fluctuating area 'Higgs field' whose quantum excitation led to the bosonic form of the field. In the Higgs mechanism, the gauge fixing imposed, which is regarding the stabilization of distinction of degrees of freedom at a minimal extremity in the spontaneous symmetry breaking and this whole concept is called the Unitary gauge; in this way the mechanism has primordial transformation to bosons, which is an underlying exigency to the spontaneous disintegration of incessant symmetry and to be precise, this is called Nambu—Goldstone bosons. There are a few things which are applicable to both the field theories of strong and weak interactions, for example the Fermi Liquid Theory, flavour symmetry which has a subset known as Isospin, etc. But there are a few things that are poles apart, for example there is an elementary particle of half integer spin which has no interface in the realm of strong interactions but plays a significant role in all other fundamental interactions called leptons. Because these points are more concerned with the world of particle physics, I will rest the explanation of these interactions at this point only, but assuredly all these abstracts have determined various new facets of sciences in the last few decades and hence, particle physics is also attributed to quantum physics only in some sense.

31

Quantum Topology &The Topological Quantum Field Theory

Henceforth space by itself, and time by itself, are doomed to fade away into mere shadows, and only a kind of union of the two will preserve an independent reality.

– Hermann Minkowski

From last few decades, pure mathematics has been applied intensively in quantum physics, although in other aspects of theoretical physics, we had certain notions of differential geometry right there from the time of Albert Einstein. But at this point of time, differential geometry is playing a significant role in a specific side of quantum physics as well. Before knowing the core relevant abstracts, one should be familiar with some technical terms that uses in the unification of quantum physics and differential geometry. First of all, differential geometry itself is the imposition of certain mathematical terminologies like calculus, algebraic geometry, etc. in order to deal with geometric interpretations of different degrees and dimensions; thus it has some very distinguished applications in different fields of mathematics and physics. General Relativity is the best example of it. Einstein combined with Hermann Minkowski are the scientists so far who have intended utilizing differential geometry for describing the reality, but specifically Einstein who brilliantly brought it into the play of nature through his Field Equations and after his insights, a lot of development has been made in this area which led to various magnificent discoveries such as in geometrothermodynamics (GTD), electromagnetism etc. There are actually several different subsets to differential geometry and most of the branches come under different forms of Topology, which is the study of continuous geometric transformations in terms of some respective intrinsic properties. But differential geometry has become a salient feature of quantum physics just a few decennials back. Although a few physicists and mathematicians of early times had big hand in the development of this set of mathematical tools, but for the current course of time, we will take the legacy of the remarkable mathematician Leonhard Euler for his development of a kind of invariant form of topology by the use of his famous and fundamental analogy of vertices, faces and edges; all these three entities remain as it is in terms of different shapes no matter how you calculate. This conceptualization of Euler is very elementary but his insights influenced two modern mathematicians named James Harris Simons and Shiing—Shen Chern, such that around 1974, they collaborated and came up with the idea of development in topological invariants with the paper titled "Characteristic Forms and Geometric Invariants". They fundamentally used some corresponding fields and other mathematical arrangements like action of principal vector bundles and homomorphism that deals with the relation in betwixt of same set of algebraic structures or spaces in the well-known Chern—Weil theory, introduced by collaboration of Shiing—Shen Chern and André Weil, which imposes certain vector forms of topological invariants on connective mapping of smooth manifolds and is also responsible for the Chern—Simons theory. The main idea of the theory was to establish combinatorial approach to the vector bundles as real valued, i.e. 'Pontryagin classes' of the four-dimensional smooth manifold. Hence the theory ultimately developed invariance in the low-dimensional topology via some sort of absolute differential characters.

The theory was not anyway intentioned for any implications to theoretical physics and had nothing to do with any facet of physics for about ten years of its proposition, until when the Hamiltonian of mathematical representations by Chern and Simons were observed by several physicists. They found that the theory was also applicable to the quantum realm via certain gauge groups up to specific dimensions, and this was the birth of Topological Quantum Field Theory (TQFT) and Quantum Topology. The later development in TQFTs is regarding to the quantum gravity, one-dimensional complex manifold, i.e. Riemannian surfaces, Minkowski space-time, etc. which designs the topological invariants in trivial ways, that leads to the low-dimensional analysis in it, such that the higher-dimensional perceptions are still under the development. The inverse mapping of identical spaces or structures i.e. 'isomorphism' also plays a prime role in this insights which is precisely known as the 'Moduli Spaces' that represent both algebraic and geometric interpretations to these classes of isomorphism that exhibits supersymmetric gauge formulation; super-symmetry briefly is something related to bosons and fermions, thus their respective degrees of freedom with respect to corresponding fields that also plays a crucial role in discrete levels of QFTs. The specific attribution of this whole idea is nothing but the action of low-dimensional physical aspects of discrete and continuous spaces. The moduli spaces also used by many another mathematician in order to formalizing some of the well-known primitive notions for the advancements in smooth topological four-dimensional manifolds, technically by the term called as 'Anti-self-dual connections' that underlies a few techniques such as eigenvalues and differential forms with a finite-dimensional vector space in the domain of linear mapping which we call 'Hodge star operator'. Both the sides of physics and mathematics concerning with these aforementioned elements are somehow intrinsically connected as obvious, that's why the invariants generated by both verge of theories give rise to almost similar results in four-dimensional smooth manifolds. In both quantum topology and TQFT, representation theory which is related to the vector spaces undergoing linear transformations, gives an extremely noteworthy contribution; it also applies to some other quantum realm too that are closely related to a few particular things in quantum topology, for e.g. non-commutative abstract algebra or quantum group, simultaneous inclination of both algebra and co-algebra, i.e. Hopf algebra, killing vector field that give rise to the continuous isometrics and conserves metric while imposing on Riemannian manifold, etc. The mathematics of quantum topology also uses one of the most unique legacies by Paul Dirac and that is a tool or operator which generalizes the complex Hilbert spaces as linear functional vector form – Dirac notation or 'Bra-ket notation'; this is also used in several distinctive ways in quantum mechanics like wave-function, superposition, EPR Paradox, etc. TQFT and quantum topology at low-energy also has some foremost applications in condensed matter physics and statistical mechanics, such that one of the most rudimentary theory is topological degeneracy in, which the Hamiltonian of many-body system exhibits the disintegration of evolution in the macroscopic limits of ground state as the finite gap to the excitation state and has an aftermath of some sort of zero temperature order as critical extremity, i.e. Topological order, and this is also applies in the study of interaction of quantum spin and magnetic material called Quantum spin liquid. Similarly, it also has a profound application in the strong magnetic field with low-temperature state regarding the two-dimensional system, i.e. Quantum Hall effect; the preserved symmetry in topological order applies to the notion of Topological insulator. Both the fractional quantum hall effect and the two-dimensional analogy have a vital application in what we call the Topological Quantum Computer, which is a theoretical quantum computer that governs over "anyon", which is a kind of quasi-particle. In statistical mechanics, especially in one-dimensional system that leads to the distinction of inner quantum states as conservation of momentum, topological techniques plays a central role and this is known as Yang–

Baxter equation. The model of statistical mechanics in lattice form, which is known as Classical rotor model, is also related to the condensed matter physics and the two-dimensional phase transition as any quantized flux that is topologically in a stable array i.e. 'quantum vortex' and this is called the Berezinskii–Kosterlitz–Thouless transition or BKT transition under what specifically called as the circular symmetry. The effective study on fractional quantum Hall effect and BKT transition, which are one of the most primitive developments of quantum topology in condensed matter physics in combined form brought a Nobel Prize of Physics in 2016 to David James Thouless, John Michael Kosterlitz and Frederick Duncan Michael Haldane, which was officially for "theoretical discoveries of topological phase transitions and topological phases of matter". However, the physicist Vadim Berezinskii whose work led to the potential discovery was died earlier in 1980.

There are some more formulations in quantum topology, which have their own special peculiarities like Cohomology, comes under the domain of algebraic topology, i.e. topological spaces are associated with the abelian groups, although in mathematical terms homology is also similar to it but cohomolgy has more profound invariant aspects when concerned with the corresponding spaces, for e.g. rings, lattices, etc. But the Symplectic Geometry, which is a branch of topology, arose from the classical Hamiltonian mechanics and only specifically deals with the two-dimensional closed form in differentiable spaces as non-singular form which exhibits isomorphism as a bilinear mapping and the consequence of this framework is what we are looking for, i.e. quantum cohomology. Another framework known as the Knot theory, which is a mathematical analysis of knots within the range of Euclidean space embedding, or three-dimensional to be precise, that only has continual deformations as underlying property and this study has a big hand in the notion of Jones polynomial discovered by mathematician Vaughan Frederick Randal Jones; these polynomials determine rational orientations and other related aspects like invariance of the knots, but when it comes to quantum topology or TQFT, it acts with Chern—Simons theory but doesn't formulate with three dimensions of space and one dimension of time, rather deals with two dimensions of space and one dimension of time. Thus, it doesn't generalize the knot to the mathematics of four fundamental dimensions, although many distinguished complex form centric homology later extended the space-time analogy of Jones polynomial for knots to the four-dimensional analysis, and up to the current time, it is remained a subject of research regarding the generalization of more higher-dimensions. There are some more group of invariants in quantum topology that are focused on the knots for three-dimensional compact orientate manifolds whose conventional impacts were enhanced by several mathematicians called with various mathematical terms of invariants in different realms of abstract geometries. In the symplectic topology, these invariants particularly plays as crucial aspects which have an underlying condition of Pseudo-holomorphic curves that satisfies the set of two partial differential equations as complex differentiable functions, i.e. Cauchy—Riemann equations while imposing smooth mapping from Riemannian surface. The Yang–Baxter equation mentioned earlier is also included in knot theory via corresponding groups, i.e. Braid Group, and this is applied in the notion of Topological entropy.

Such that, the rise of quantum topology and the TQFT is integrated with plenty of aspects in sciences, in the form of some very elegant and precise mathematical structures and tools.

Part 8:
The Probabilistic Future with Quantum Physics

32
Core Future Possibilities In Theoretical Sciences

There is geometry in the humming of the strings, there is music in the spacing of the spheres. - Pythagoras

Theoretical sciences are the pillars of scientific perspectives, and in physics we have two very elegant and precise notions as the descriptions of reality - Einstein's Theory of General Relativity and quantum physics. So somehow the greatest chances of becoming the next milestones lies in these two only. But before knowing the ultimate majors and minors in these areas, one has to be familiar with some keywords which are very fundamental in the study of advanced versions of these two descriptions and these are – unified field theory, quantum gravity and string theory. The keywords are some sort of prerequisites for the upcoming noted mentions. However, all of these terms came into existence post either general relativity or quantum physics, or both. One would think that how both can act simultaneously knowing that their respective explanations for the nature of reality are poles apart in some sense, but it's true that the above mentioned keywords consist of terminologies from both the explanations somehow and because sciences are seeking the ultimate truth as a vision of the great Albert Einstein, that's why their unification has been a necessary task to develop and the origin of this idea was also Einstein's in some ways.

As we have known from previous chapters that Albert Einstein with his general theory of relativity described the wrap and curve in the fabric of space-time in 1915, and subsequently in 1919 the astronomical observations favoured the theoretical and mathematical representations of Einstein and the observations were mainly attributed to the great astronomer Sir Arthur Stanley Eddington. But Einstein had another idea, with respect to the unification of all the fundamental forces of nature, i.e. electromagnetic and gravitational forces, as both were actively known at that time and this terminology is called the Unified Field Theory, which was coined by Einstein himself, defined in one of the previous chapters concerning Schrödinger's affine field theory. Although in particle physics, there is a similar kind of model, which is called the Grand Unified

Field Theory that deals with three fundamental interactions simultaneously as mentioned in one of the previous sections covering electroweak interaction, but this version of unification has some limitations and depends on various conditions and gauge groups and is not suitable to our current topic. So, let's continue with the classical approaches to unified field theory.

Another brilliant German physicist and mathematician, Theodor Franz Eduard Kaluza had some parallel insights to the unification phenomenon. Hence he thought to try the same kind of technique that exhibits wrap and curve with any other primordial force, he chose the electromagnetic one like Einstein had chosen gravity. But then an obstacle arose regarding the medium of wrap and curve because Einstein had already used the space-time notion of the four dimensions in his general theory of relativity or field equations of gravity. But then Kaluza got a very bizarre idea, which was about a possibility of some other additional dimensions that can deal with the phenomena he was seeking, a medium for wrap and curve to generalize the field equations for electromagnetism. Then he went on to define the respective field equations for electromagnetism with four dimensions of space and one for time. The elegant mathematical framework turned out to be brilliant in two different forms as identical manifestations which are mentioned as follow:

1. The field equations were extension to the Einstein's Theory of General Relativity to five dimensions.

2. Because the equations had something to do with both general relativity and electrodynamics, thus it

became a significant step towards the unified field theory.

Now, we have got two very marvellous conclusions/ideologies from the theory of Kaluza on electromagnetic phenomena but, that's not the end to the story because there are two consequential fundamental questions that arose around extra-dimensions and the questions were :

1. If there are any extra dimensions then why we can't see them or where are they?

2. Does this theory have something to do with the reality of nature?

The first question was answered around 1926 when a physicists Oskar Benjamin Klein came up with his opinion on Kaluza's theory and according to Klein, there is not any single conceptualization of dimensions as they are of two distinctive types; the first type is what we can see or feel, i.e. macroscopic dimensions and the other ones are very subtle and curled up onto the fabrication of large-scale dimensions and with this insight Klein gave quantum interpretation to the Kaluza's theory concluding that the additional fourth dimension of space is curled up with a very microscopic radius. Due to the contribution of Klein, the idea is now known as Kaluza-Klein Theory. Now, coming to the answer of second rational question, which is about the dealing of theory to reality. The answer is sadly no! because we can't define anything about the nature through this idea and due to quantum interpretation over it, later on the concept of quantum gravity was developed which laid the concept of 'graviton' as a hypothetical elementary particle that oscillates over the quantum field to gravitation and it remains a major unsolved aspect in the quantum world up to the present. There are not many clues in this area that's why only this concept hasn't been mentioned in the course of this book. But the other aspect is the notion of string theory, although in the first go, string theory had nothing to do with this concept of extra dimensions because it was developed to define the very subtle nature of entire richness as the ultimate being of any particle, like we all know now that the current ideas, which we have gone through in quantum or particle physics end up to the quarks, but there is another entity which is quite fine compared to the quarks and that set-up is a tiny vibrating strings oscillating with different vibrations and each of them as the distinction determines the nature of corresponding particle responsible for the abundance we see around us.

But here we observe the unity due to a singular entity, which is absolutely irrespective of the individuality and then the idea come into the course of action in string theory that it may be Einstein's idea of unified field theory. But the mathematics of string theory regarding the unification possesses a condition which suggests that there must be at least ten dimensions of space in order to study the oneness of all fundamental forces of nature. Although, it is said that some of such theories also suggests a hypothesis of a whole chain or group of universes out there called the 'Multiverse', but there is no rationalism as of now to support these points. Attaining the string theory can be a vital consequence as a facet of quantum physics in unison with general relativity; apart from that, the quantum gravity can also be visualized simply with the idea of unification of quantum physics and general relativity, which hasn't been accomplished up to now. In the whole scenario of quantum realm, there is a thought that gravity can't be unified with quantum physics because the force is very weak, but if we take the case of black-hole, then gravitational field is extremely intense but we don't have a singular affinity to deal with, that too with the quantum unification with gravity because perhaps there are some kind of discrepancies in our vision because general relativity and quantum physics have different kind of underlying perceptions regarding the space-time and reality, as Einstein's idea is deterministic and treats the space-time as smooth whereas, quantum physics deals with the entire phenomena as discrete quantization and superposition. Perhaps one can't even know the exact place of a black-hole and the corresponding state if it concerns the quantum explanations. Similarly the intrinsic uncertainty in the quantum mechanics may create plenty of cosmological crises in terms of many rudimentary aspects like space-time singularities and beyond. Hence, quantum mechanics has to be deterministic for the nature of reality or some modifications should be there that deals with the determinism in a precise way. Subsequently general relativity also has to be developed into a bigger regime of natural explanations, so that it can make both quantum world and the large-scale universe compatible with each other and humanity can have a better intuitive understanding to several new perspectives. So, in theoretical sciences this, alongside the aspects of Quantum Cosmology is one of the key possibilities we have in terms of quantum physics. Although, there can be a few more enhancements, like the concept of renormalization is not a theory but a set of certain rules so it can be somehow determined as theory. Apart from that, quantum applications can be there in economics too by defining some more profound formations of efficient strategies, econometric solutions, transportation theory and beyond.

33
A Futuristic Insight to The Quantum Technology

We can only see a short distance ahead but we can see plenty there that needs to be done. – Alan Turing

From last several years, specifically the modern times; humans have been able to extract plenty of technologies which have become very significant part of our daily lives due to their immense capabilities. There is no question to the fact that technology plays one of the major roles in the development of humanity if used or applied in such a way as to benefit the world. It's very obvious that sciences plays a key role in the development of a nation because everything takes place in some form of science for example, the legacy of great mathematician and cryptographer Alan Turing in terms of the mathematical perceptions in code-breaking which helped a lot to his aura to be developed in a more distinctive way. Similarly, Einstein's relativity is one of the greatest reasons behind the existence of internet or Global Positioning System (GPS). But, in the current scenario, they are classical versions in some sense while concerning with technological terminologies and that's one huge difference between theoretical sciences and technological aspects because Turing's visionary mathematics and Einstein's great framework are still astonishing and keep having various new phenomena as a consequence to deal with, but their realistic applications are need to enhanced so that we can have more and more stunning technologies and for this vision, quantum physics is the key in these times, even with its uncertain laws. Quantum forms of technologies will certainly devise some very mind-blowing automations and one such aspect which is fabulous and equally wonderful within the uncertain explanations of quantum mechanics is Quantum Computer.

Unlike the classical computers, which were developed on the basis of binary operations as fundamental units, quantum computers are based on its own units known as Qubit or Quantum bits that are specifically noted for the artificial form of quantum information. The data in quantum computers lies according to the superposition principle means if one intends to play a game in a quantum computer then the consequences will exists in probabilistic form simultaneously as precondition with respect to the number of distinctive programmed cases, but the catch is that its uncertain to us, but not for the computer; hence one cannot dominate the quantum computer because it has deterministic simulation in internal terms and that's why its programming regarding anything is not a simple task to achieve. But yes, it can perhaps be decoded with the attribution to artificially coded mathematical computations though this too is a tough task; nevertheless there is this possibility with quantum computers. It is to note that in principle, Quantum codes, if set-up precisely can provide better security to one entire technical system throughout Nations if implemented in a compatible way. Making of a quantum computer require the laws of quantum mechanical properties with a very close precision, but the most fundamental tool is superposition principle and perhaps eigenfunction too, because there must be some linear transformations imposed to attain the programming at quantum scales. The current analogy of quantum computers can possibly be integrated using the TQFT or some specific mathematical explanations in the quantum topology because they deal with several kind of transformations including advances like conformal ones, thus they may be utilized as precise tools to formulate the quantum simulations in multiple states. Quantum physics is surely capable of making great contributions in the area of medical sciences too, for example medicines have certain molecular structures which can be analysed as any predominant characteristics through quantum properties due to a common regime of artificial functional computations, if conceived in terms of quantum computing. Apart from quantum methods in technological systems, two very genuine questions arise which described below :

1. Is it necessary that quantum physics has to be deterministic like in theory for defining reality of nature?
2. Why is it, that quantum mechanics doesn't show any kind of discrepancies when applied as technology?

The answer to the first question is yes, it's true that there is no substitution to the determinisms in the theory and even in technology too, but the uncertainty of quantum mechanics makes several aspects very profound like security of encryption and so on. Now, the answer of second question is that we don't have a lot of idea about implementations of quantum technologies in the present point of time and that's why there is a possibility that it shows some discrepancies. In fact, there are pretty good chances for that to happen because quantum computing is not at all different to quantum mechanical properties and the corresponding mathematics, and we already know that they would have to be modified to deal with the reality of nature. So perhaps we will get to know about all such questions in future or in quantum future. But, perhaps we can get more precise technological aspects when the uncertainty from quantum world doesn't exist anymore and then our technological aspects will be able to give results more effectively so that it could be possible some day to even scrutinize the mysteries of this elegant universe or with quantum computing. Apart from uncertainty in quantum computing, we can make it more effective if we can determine the qubits in functional way at which, quantum topology may be a useful area to investigate on such that, in the field of computer science, quantum physics could play an important role in order to define the many fundamentals of the modern era and upcoming times.

There are some more quaint in the technologies of quantum world and one of them is simulation. Well, it's not the conventional one but a phenomenon of real world which is a possibility from quantum computers perhaps in the future. Simulation is an idea which suggests that the real universe is an operated one by some other more advanced civilization with its own laws of physics, chemistry and biology; or with other economical dynamics and so on. It can be possible with the notion of multiverse and may be in future we are able to develop a technology which could give rise to a simulation to some other universes (if they exist). Though it seems like a fantasy but this strange possibility exists. However, right now, we are unable to do such a thing but who knows if our civilization reaches level where we can attain victory and realise our imagination, because thousands or hundreds of years ago no one would have thought that one day humans would fly too or capture their present in the form of live memories. So why not this because may be there is a possibility that among all the chains of universe, i.e. multiverse; only one is real, which is not simulated and all others are just simulation of chains, just like a robotic scientist ensembles or programs one robot and other robots are programmed consecutively by robots only. So, realistic simulation may turn out as one of the degrees of quantum computing. Another facet that is quite fascinating to some people and is just seems like the EPR Paradox or Quantum entanglement i.e. "teleportation", which generally refers to two distinctive things affecting or transmit information to each other without being physically connected simply because their states are entangled as mentioned in the section of EPR Paradox, however, this term is both non-scientific and quite irrational, even to ask questions of such sort in the science of logic and intuition should be turned down sustaining the sake of rationality. Going back to the rationalism, there can also be a possibility of hidden variable theory because everything is deterministic in the universe. Quantum tunnelling, which is responsible for the flow of probabilities to two distinctive ends as the one-dimensional wave-mechanical descriptions known as rectangular potential barriers, also has some vital consequences to the quantum computing or vice-versa. Alongside, the Quantum Complexity and Knot Theoretic incorporation in it, can also be a far-fetched future to the Quantum Physical integration.

Again, Einstein comes into play because the unification of general relativity and quantum physics can lead to a more powerful and a more precise GPS compared to the current one and Bose—Einstein condensate can also lead to various advanced notions in quantum computers because it can to lead the formation of atomic lasers due to the simultaneous statistical vibrations of atoms, and this may be responsible for various striking additions and gain quantum computing stunning success. But I hope that humanity, also stays beware of such futuristic technologies. However in this way, we are seeking development as certain deep understandings and implementations of technology but as far as the universe is concerned, then even when homo-sapiens become capable of exploring the reality using any kind of advanced computations, I'm not convinced anyway that they can ever decode the ultimate mystery of the Universe.

Acknowledgements

Writing the book; especially for the general audience is one of the key morals I came across. The reasons behind this entirety of thought are for scientific, philosophical and rational awareness among people and as tribute to Albert Einstein. Since, this technical literature is brought out for general audience, therefore there were a few places were scientific and mathematical technicalities are intentionally compromised to make them accessible for non-technical audience.

Creativity is the only key which drives the nature and writes history but at the same time, those who understand that drew originality are also intrinsically determined in that part of historic elements. I remember my first meeting with Mr. Deepak Ojha a few years back, the Managing Director of the esteemed corporate the "Waldies Compound Limited". The meeting held to be very long & intensively resolved such that I saw a distinctive fortitude in his capabilities as the carried inner dexterity towards archetypal intellect which makes him as an intellectually autochthonous person from core. The discussion went with a range of diversified ideas, however its extremity was pointed onto the concept of this book and while understanding the potential of its capturing in the direction of scientific ignition in general audience, he immediately proposed to manage for its production and post-production association. Artist forms an art but that peculiarity of peace is worthwhile to the masses when someone like Mr. Deepak comes along.

Simultaneously, there is something irrefutable here which deserves to be at the centre regime. I'm much gratified towards Mr. Sushil Ojha, the Founder of widespread humanitarian organizations with amalgamation of humanity & crystallized socialism. Mr. Sushil, in combination with the asceticism of Mr. Deepak has put forward a whole team and management to look after the progression of the book on each single point throughout the publication process. It was my first interaction with Mr. Sushil but it made me flabbergasted to see how his allegiance as truly perfervid towards the integration of society in such a way that everyone get inflamed for a better sight of efficacious variance towards the place & environment we live in. He has staggeringly developed a milieu of native humanism through his promises to the society which will be sustaining many lives to be galvanized for the identical facet of bequest.

Alongside, I'm thankful to all those esteemed people who managed to meet my ideas at very early extremity and subsequently vitalized them to be integrated across the disciplines, it will be difficult to name all of them but primordially includes Mr. Pradeep Kumar Sinha (Economist, former Cabinet Secretary of the Republic of India), Dr. (Prof.) Manjul Bhargava - Fellow of the Royal Society (FRS) (Fields Medallist, Mathematician - Princeton University, USA); Dr. Krishnaswamy VijayRaghavan - Fellow of the Royal Society (FRS) (Former Principal Scientific Adviser to the Government of India); Dr. (Prof.) Umesh Garg (Physicist & Director – REU, Notre Dame University, USA; Fellow of American Physical Society & American Association for Advancement of Science; USA); Dr. Pankaj S. Joshi (TWAS Fellow, Founding Director –

International Centre for Cosmology & Ex- Senior Professor in Dept. of Astronomy and Astrophysics – Tata Institute of Fundamental Research, Mumbai) and several others.

The decision to draw upon this rational canvas, was enormously encouraged and morally endorsed by a few highly major people that intrinsically involved in my life; in this realm my parents Mr. Shelender Kumar Panchariya and Mrs. Anju Panchariya, comes at the foremost place due to their immense love and support throughout the journey. The conclusion of this literature makes me heartfelt emotive, since with this; I also endeavour to have sustained the wishes and blessings of my grandfather Mr. Liladhar Panchariya, to contribute in the quest of wisdom and beyond.

Although, I've constraints of words but even with this clampdown; due to their cordial affection, I'm henceforth thankful & gratified to all staunches held by these incredibly colossal people to have faith in pursuit of the purest form of art which is portraying & diversifying the continuous discreetness of Universe.

Dev Arastu Panchariya

Author

Quantum Physics being one of the most integral & complex chunk of the overall sciences, from a long history; great minds like Shri. Dev Arastu Panchaiya have taken the all important mantle of making this field more viable and comprehendible for the scientific brains who are scattered around the planet.

His inspiring and magnetic persona as a Great Polymath expresses a sincere depiction of what a gifted as well as sedulous man Mr. Panchariya is.

Being in close contact with VIPRA-Foundation, we well and truly know how humble, balanced & down to earth this legendary persona is, not only as a scientist and philosopher with abundance of clinical knowledge but also as a human being with sterling nature and deep wisdom.

We, as well wishers of Mr. Panchariya are tremendously happy and proud for all the titles on which he has his name carved. Subsequently, we wish him all the very best from the bottom of our hearts for his book on Quantum Physics which by all means can revolutionize the entire spectrum of this particular subject for the greater good. He holds a success, he truly deserved. It is an achievement he has truly earned which is inspirational for society.

Entire VIPRA-Foundation would like to wish Mr. Dev Arastu Panchariya all the luck for his future endeavors.

Sushil Ojha

Founder - VIPRA Foundation

Humans have always been fascinated by the Nature and its phenomenon. The curiosity to know more and more has been pivotal to the transformations that took place ever since the stone-age. During this long period of time to the present electronic-era; at instances, Nature had selectively decided, through whom it is going to unfold its mystery. It gives me immense pleasure as well as pride to experience such a nature's selection in Dev Arastu. His sound understanding and enormous wisdom in the diverse range of areas at such young times is not less than any miracle. I have a strong belief that his book "Einstein & the Evolution of Quantum Physics: From light beam to deep cosmos" would give readers an entirely new perspective to look at the quantum framework.

From the history of science, we know that the development of classical mechanics that took place over the period of about a century, laid the firm foundation for the Quantum Mechanics. Subsequently, as all the laws were redefined in accordance with Quantum Mechanics, it enabled a paramount development in the last century. Now, after a century of the significant developments in the field of Quantum Mechanics, a new formalism to cater unsolved problems of present is much expected. I would not be surprised if this book serves the purpose of igniting readers mind and may prove to be revolutionary for the development of something entirely novel. I extend my congratulations to Dev Arastu for bringing his intuitive thoughts in the form of amazing text in this book and best wishes to the readers.

Pankaj Ojha
Director,
Deutsche Bank, Florida
United States of America

It has always been a pleasure and exciting expedition talking to the hyper-intellectual Dev Arastu over the past few years. I have worked with countless scientists from different areas in academic interactions, but his determination and resolve are really very inspirational and truly exceptional. I congratulate him for this unique endeavour of writing a book on a very relevant subject, especially at this very young age.

I am really overwhelmed by his enthusiasm that has guided him to write a book on science pertaining to his own acumen on research and knowledge. In science, there is no "ultimate truth" or "onetime truth", scientific truth is always evolving. This is the reason that his wisdom and determination taking him to greater heights.

I really hope that the new generation students, scientists and researchers will ask more questions and greatly benefit from his book on "Quantum Physics".

I am optimistically looking forward to this spectacular book stimulating people for the dignity of sciences.

Dr. Madhuchanda Kar

Ex-Chairperson, BOG, IISER-Mohali

Governing Council Member (Ministry of Education Nominee),

Central University – Gaya, South Bihar

Clinical Director, Dept. of Oncology

Peerless Hospital, Kolkata